# RADIATION PROTECTION

## ICRP PUBLICATION 30

### PART 2

# Limits for Intakes of Radionuclides by Workers

A report of Committee 2 of the
International Commission on Radiological Protection

ADOPTED BY THE COMMISSION IN JULY 1978
This report supersedes ICRP publication 2

PUBLISHED FOR

The International Commission on Radiological Protection

Los Angeles | London | New Delhi
Singapore | Washington DC

# CONTENTS

# Acknowledgements

The following were the members of ICRP Committee 2 who prepared this report.

J. Vennart, (Chairman); W. J. Bair; G. W. Dolphin (Secretary—died August 1979); L. E. Feinendegen; Mary R. Ford; A. Kaul; C. W. Mays; J. C. Nenot; B. Nosslin; P. V. Ramzaev; C. R. Richmond; R. C. Thompson and N. Veall.

The committee wishes to record its appreciation of the substantial amount of work undertaken by N. Adams and M. C. Thorne in the collection of data and preparation of this report, and, also, to thank P. E. Morrow, a former member of the committee, for his review of the data on inhaled radionuclides, and Lorene Ashby and Joan Rowley for secretarial assistance.

The dosimetric calculations were undertaken by a task group as follows:

Mary R. Ford (Chairwoman); J. R. Bernard; L. T. Dillman; J. W. Poston and Sarah B. Watson.

Committee 2 wishes to record its indebtedness to the task group for the completion of this exacting task.

# APPRECIATION

## G. W. DOLPHIN

Geoff Dolphin was a member of ICRP Committee 2 from 1969 and its Secretary from 1973 until his untimely death on 20 August 1979. He held the office of Secretary with distinction and through it brought an order to our business that will be difficult to emulate. Everyone on the committee appreciated his non-conformist views and his good common sense.

He enjoyed the good things in life and the company of friends he had made in many countries. After he became Secretary, being a member of Committee 2 took on a new dimension; the hard work of the day would be rewarded by arrangements he had made for a social evening and he liked nothing better than that this should be at his home with his wife Yvonne.

We were all concerned to see his health declining as he worked on, to within days of his entering hospital to seek medical advice. He will be greatly missed.

# PREFACE

The data given in this report are to be used together with the text and dosimetric models described in Part 1 of ICRP Publication 30[1]; the chapters referred to in this preface relate to that report.

Since the publication of Part 1 the Commission has decided, as reported at the front of this volume, to reduce its recommended dose equivalent limit for the lens of the eye from 0.3 Sv in a year to 0.15 Sv in a year. The new value is used here when considering Derived Air Concentrations (DAC) for radioisotopes of the noble gases argon and xenon. Consequential changes in the values of DAC given in Part 1 for some radioisotopes of krypton are reported in the Addendum (at the end of this report).

To derive values of the Annual Limit on Intake (ALI) for some radioisotopes of the elements considered here, the following additions to the text of Part 1 have been used.

## 1. DAC for Submersion in $^{37}Ar$

The methods discussed in section 8.2.3, Chapter 8 are valid as stated for all the radioisotopes of the noble gases, argon, krypton and xenon that emit either photons or $\beta$-particles of considerable energy. However, $^{37}Ar$ decays by electron capture emitting Auger electrons of insufficient energy to penetrate the skin to the basal layer of the epidermis at a depth of 70 $\mu$m and X-rays of energy less than 3 keV of which only a small fraction will penetrate to this depth. In these circumstances it is appropriate to limit exposure to $^{37}Ar$ in the manner described for elemental tritium in section 8.2.1, Chapter 8 by consideration of the dose-equivalent rate in lung.

## 2. Absorbed Fractions

In the metabolic data for indium, part of the element leaving the transfer compartment is translocated to red marrow, which is not considered as a source organ in Chapters 4 and 7. For that part of the radioisotopes of indium in the body which is in red marrow the absorbed fraction for photons is taken from Snyder et al.[2] For $\beta$-particles from radioisotopes of indium in red marrow the absorbed fraction in red marrow, AF(RM $\leftarrow$ RM), is assumed to be 1. The dose equivalent in bone surfaces adjacent to red marrow is taken to be the same as the average dose in red marrow so that the dose equivalent to the whole of bone surfaces, BS (of which only half is adjacent to red marrow, section 7.2, Chapter 7) is one half that in red marrow.

For radioisotopes of gold in the bladder the dose equivalent from $\beta$-particles in the bladder wall is taken to be half the dose equivalent in the bladder contents. Values of absorbed fraction for photons are taken from Deus et al.[3].

Both technetium and rhenium are translocated from the transfer compartment to stomach wall, which is not considered as a source organ in Chapter 4. For radioisotopes of these elements the absorbed fraction for $\beta$-particles AF(ST wall $\leftarrow$ ST wall) is taken to be 1, and for photons values are taken from Deus et al.[3]

Brain is a source organ for both copper and mercury. Values of absorbed fraction for photons, AF(Brain$\leftarrow$Brain), for radioisotopes of these elements have been computed by methods described by Snyder et al.[4]; for $\beta$-particles AF is taken to be 1.

In the following pages the relevant metabolic data for each element precede a table of values of ALI and DAC for radioisotopes of that element having radioactive half-lives greater than 10 min.

The metabolic models described are for compounds of a stable isotope of the element.

Retention data given in the literature have, where necessary, been corrected for radioactive decay of the radionuclides concerned. Because of the considerable variation of gastrointestinal absorption from individual to individual, values of $f_1$, the fraction of a stable element reaching body fluids after its entry into the gastrointestinal tract, are given to one significant figure only (Section 6.2, Chapter 6). For inhalation, values of ALI and DAC are given for each different inhalation class (D, W and Y) appropriate for various compounds of the element (Chapter 5).

In the metabolic data, when the symbol $t$ is used for time its unit is always the day unless specified otherwise.

Values of ALI (Bq) are given for the oral and inhalation routes of entry into the body. It is emphasized that the limit for inhalation is the appropriate ALI and that the values of DAC ($Bq/m^3$) for a 40-h working week are given only for convenience and should always be used with caution (Section 3.4, Chapter 3). Values of ALI for inhalation and DAC are for particles with an AMAD of 1 $\mu$m. A method of correcting the values for particles of other sizes is described in Section 5.5, Chapter 5 and the required numerical data are given in the Supplement to this Part.

If a value of ALI is determined by the non-stochastic limit on dose equivalent to a particular organ or tissue, the greatest value of the annual intake that satisfies the Commission's recommendation for limiting stochastic effects is shown in parentheses beneath the ALI. The organ or tissue to which the non-stochastic limit applies is shown below these two values. When an ALI is determined by the stochastic limit this value alone is given (Section 4.7, Chapter 4).

All the values of ALI and DAC given here are for occupationally exposed adults and must be used with circumspection for any other purpose (Chapter 9).

## References

1. *ICRP Publication 30*, Part 1, *Limits for Intakes of Radionuclides by Workers.* Annals of ICRP 2, No. 3/4, 1979.
2. Snyder, W. S., Ford, Mary R. and Warner, G. G. Estimates of specific absorbed fractions for photon sources uniformly distributed in various organs of a heterogeneous phantom. MIRD Pamphlet No. 5 Revised, Society of Nuclear Medicine (1978).
3. Deus, S. F., Provenzano, V. and Snyder, W. S. (1977). Specific absorbed fractions for photons emitted in the walls of the GI tract. *Health Phys.*, **33**, 191–197.
4. Snyder, W. S., Ford, M. R., Warner, G. G. and Watson, S. B. (1974). A tabulation of dose-equivalent for microcurie-day for source and target organs of an adult for various radionuclides, Oak Ridge National Laboratory Report ORNL-5000.

# METABOLIC DATA FOR FLUORINE

## 1. Metabolism

Data from Reference Man[1]

| | |
|---|---|
| Fluorine content of the body | 2.6 g |
| of the bone | 2.5 g |
| Daily intake in food and fluids | 1.8 mg |

## 2. Metabolic Model

### (a) *Uptake to blood*

Absorption of fluoride present in food or as added fluoride in drinking water is rapid and almost complete.[1] This seems also to be true for most inorganic compounds of fluorine[2] in solution and in this report $f_1$ is taken to be 1 for all compounds of fluorine.

### (b) *Inhalation classes*

The ICRP task Group on Lung Dynamics[3] assigns fluorides of the various elements to inhalation classes D, W or Y. For information concerning the inhalation class to be associated with the fluoride of a particular element the metabolic data for that element, or the task group report should be consulted.

| Inhalation Class | $f_1$ |
|---|---|
| D | 1 |
| W | 1 |
| Y | 1 |

### (c) *Distribution and retention*

Fluorine entering the blood is very rapidly deposited in mineral bone[4-7] and by 20. min post-injection deposition is essentially complete.[5,6]

In this report it is assumed that all fluorine entering the transfer compartment is instantaneously translocated to the skeleton. Since none of the isotopes of fluorine considered in this report has a radioactive half-life of greater than 120 min, it is appropriate, for the purposes of radiological protection, to assume that fluorine deposited in the skeleton is retained there indefinitely.[5,6]

## 3. Classification of Isotopes for Bone Dosimetry

Since none of the isotopes of fluorine considered in this report has a radioactive half-life of greater than 120 min, it is assumed that fluorine is uniformly distributed over bone surfaces at all times following its deposition in the skeleton.

## References

1. *ICRP Publication 23, Report of the ICRP Task Group on Reference Man.* Pergamon Press, Oxford, 1975.
2. Underwood, E. J. Trace Elements in Human and Animal Nutrition. Academic Press, London, 1971, pp. 369–406.
3. ICRP Task Group on Lung Dynamics (1966). Deposition and retention models for internal dosimetry of the human respiratory tract. *Health Phys.* **12,** 173–207.

4. Wootton, R. (1974). The Single Passage Extraction of $^{18}$F in Rabbit Bone. *Clinical Science and Molecular Medicine*, **47**, 73–77.
5. Costeas, R., Woodard, H. Q. and Laughlin, J. S. (1970). Depletion of $^{18}$F from Blood Flowing Through Bone, *J. Nucl. Med.* **11**, 43–45.
6. Humphreys, E. R., Fisher, G. and Thorne, M. C. (1977). The Measurement of Blood Flow in Mouse Femur and its Correlation with $^{239}$Pu Deposition. *Calcif. Tiss. Res.* **23**, 141–145.
7. Hall, L. L., Kilpper, R. W., Smith, F. A., Morken, D. A. and Hodge, H. C. (1977). Kinetic model of fluoride metabolism in the rabbit. *Environ. res.* **14**, 285–302.

Annual limits on intake, ALI(Bq) and derived air concentrations, DAC(Bq/m$^3$) (40h/wk) for isotopes of fluorine

| Radionuclide | | Oral | Inhalation | | |
|---|---|---|---|---|---|
| | | | Class D | Class W | Class Y |
| | | $f_1 = 1$ | $f_1 = 1$ | $f_1 = 1$ | $f_1 = 1$ |
| $^{18}$F | ALI | $2 \times 10^9$ $(2 \times 10^9)$ ST Wall | $3 \times 10^9$ | $3 \times 10^9$ | $3 \times 10^9$ |
| | DAC | — | $1 \times 10^6$ | $1 \times 10^6$ | $1 \times 10^6$ |

# METABOLIC DATA FOR SODIUM

## 1. Metabolism

Data from Reference Man[1]

| | |
|---|---|
| Sodium content of the body | 100 g |
| of soft tissue | 68 g |
| of the skeleton | 32 g |
| Daily intake in food and fluids | 4.4 g |

## 2. Metabolic Model

(a) *Uptake to blood*

Virtually all sodium is absorbed from the gastrointestinal tract of man[2] and $f_1$ is therefore taken to be 1.

(b) *Inhalation classes*

The ICRP Task Group on Lung Dynamics[3] has assigned all compounds of sodium to inhalation class D and this classification is adopted here.

| Inhalation Class | $f_1$ |
|---|---|
| D | 1 |
| W | — |
| Y | — |

(c) *Distribution and retention*

The biological half-life of sodium in the body is strongly influenced by the level of stable sodium in the diet, decreasing from 335 days for a daily intake of 0.25 g to 5 days for a daily intake of 30 g.[4] Studies on 3 human subjects indicated 3 components of sodium retention with biological half-lives of 8.5, 13.5 and 445 days respectively.[5] In these cases the long term component was associated with less than 0.03% of the administered sodium. Vennart[6] has also reported a long term component of sodium retention with biological half-life of 1100 days and associated with 0.35% of the administered radionuclide.

In this report it is assumed that the fraction of sodium leaving the transfer compartment which is translocated to the skeleton is 0.3. The remainder of sodium leaving the transfer compartment is assumed to be uniformly distributed throughout all other organs and tissues of the body. Of sodium deposited in the skeleton fractions 0.99 and 0.01 are assumed to be retained with biological half-lives of 10 days and 500 days respectively. Sodium deposited in any organ or tissue of the body other than the skeleton is assumed to be retained with a biological half-life of 10 days. These retention functions are consistent with the observed retention of sodium in man and with the organ and tissue contents of stable sodium given for Reference Man.[1]

## 3. Classification of Isotopes for Bone Dosimetry

Since sodium is mainly associated with intra- and extra-cellular fluids, radioactive isotopes of sodium are assumed to be uniformly distributed throughout mineral bone, bone marrow, skeletal cartilage and periarticular tissue at all times following their deposition in the skeleton.

### References

1. *ICRP Publication 23, Report of the ICRP Task Group on Reference Man.* Pergamon Press, Oxford, 1975.
2. Wiseman, G. Absorption from the Intestine. Academic Press, London, 1964, pp 191–198.
3. ICRP Task Group on Lung Dynamics (1966). Deposition and retention models for internal dosimetry of the human respiratory tract. *Health Phys.,* **12,** 173–207.
4. Smiley, M. G., Dahl, L. K., Spraragen, S. C., Silver, L. (1961). Isotopic Sodium Turnover Studies in Man: Evidence of Minimal Sodium (Na$^{22}$) Retention 6 to 11 Months after Administration. *J. Lab. Clin. Med.* **58,** 60–66.
5. Richmond, C. R. and Furchner, J. E. Estimation of radiation protection guides: Interspecies correlations. In: *Proc. of the First Inter. Congr. of Rad. Prot.,* Rome, Italy (September 5–10 1966) Vol. 11, Pergamon Press, New York (1968), pp. 1417–1431.
6. Vennart, J. External Counting. In: *Diagnosis and Treatment of Radioactive Poisoning,* ST1/PUB65 (IAEA, 1963) pp. 3–22.

Annual limits on intake, ALI(Bq) and derived air concentrations, DAC(Bq/m$^3$)
(40 h/wk) for isotopes of sodium

| Radionuclide | | Oral | Inhalation Class D |
|---|---|---|---|
| | | $f_1 = 1$ | $f_1 = 1$ |
| $^{22}$Na | ALI | $2 \times 10^7$ | $2 \times 10^7$ |
| | DAC | — | $1 \times 10^4$ |
| $^{24}$Na | ALI | $1 \times 10^8$ | $2 \times 10^8$ |
| | DAC | — | $8 \times 10^4$ |

# METABOLIC DATA FOR SULPHUR

## 1. Metabolism

Data from Reference Man[1]

| | |
|---|---|
| Sulphur content of the whole body | 140 g |
| of soft tissues | 120 g |
| Daily intake in food and fluids | 0.85 g |

## 2. Metabolic Model

(a) *Uptake to blood*

The fractional absorption of sulphur from the gastrointestinal tract of man is typically greater than 0.6 for organic compounds of the element.[2,3] From the experiments of Dziewiatkowski[4] the fractional absorption of $^{35}S$ ingested as $Na_2{}^{35}SO_4$ can be estimated to be greater than 0.75. Elemental sulphur is less well absorbed from the gastrointestinal tract than are inorganic compounds of the element.[5] In this report $f_1$ is taken as 0.8 for all inorganic compounds of sulphur and as 0.1 for sulphur in its elemental form.

(b) *Inhalation classes*

The rate of clearance of radioactive sulphur from the lung depends very much upon the compound of the element which is inhaled. The ICRP Task Group on Lung Dynamics[6] has assigned sulphates and sulphides of all elements to either inhalation class D or inhalation class W.

For information concerning classification of sulphates and sulphides of a particular element the metabolic data for that element, or the Task Group Report[6], should be consulted.

Elemental sulphur is assumed to be in inhalation Class W.

For the purposes of radiological protection it is assumed that $f_1$ is 0.8 for elemental sulphur and all inorganic compounds of the element entering the gastrointestinal tract following inhalation, although it is acknowledged that this is probably a rather conservative assumption for elemental sulphur.

| Inhalation Class | $f_1$ |
|---|---|
| D | 0.8 |
| W | 0.8 |
| Y | — |

The inhalation of the gases $SO_2$, COS, $H_2S$ and $CS_2$ must also be considered. From a review of the literature Vennart and Ash[7] have concluded that it would be prudent to assume that sulphur inhaled in these forms is quantitatively taken up into the blood stream.

In this report it is assumed that sulphur entering the lungs in the form of $SO_2$, COS, $H_2S$ or $CS_2$ is completely and instantaneously translocated to the transfer compartment and that from then on its metabolism is the same as that of sulphur entering the transfer compartment following ingestion or inhalation of any other inorganic compound of the element.

(c) *Distribution and retention*

Experiments on rats[3] and experience in man[8-11] indicate at least two components of retention. In man a rapid phase of clearance with a biological half-life of 0.3 days is observed.[8-10] This is followed by a slower phase of clearance with a biological half-life of at least 7 days[9,10] and possibly as much as 80 days.[11] However, two such components are insufficient to explain the total body content of sulphur given for Reference Man[1] and a third, long-term, component must be postulated.

At early times after injection sulphur is fairly uniformly distributed throughout the various organs and tissues of the body[12] and a similar distribution is observed for stable sulphur.[1] In this report it is assumed that of sulphur leaving the transfer compartment fractions 0.15 and 0.05 are distributed uniformly throughout all organs and tissues of the body where they are retained with biological half-lives of 20 and 2000 days respectively. The remaining fraction of sulphur leaving the transfer compartment is assumed to go directly to excreta.

(d) *Organic compounds*

The metabolic behaviour of organic compounds of sulphur differs considerably from the metabolic behaviour of inorganic compounds of the element. Organic compounds of sulphur such as cystine and methionine become incorporated into various metabolites.[5] Thus, sulphur entering the body as an organic compound is often tenaciously retained.[3] The metabolism of organic compounds of sulphur is not considered further in this report and no limits on exposure are given for these compounds.

### 3. Classification of Isotopes for Bone Dosimetry

Sulphur is assumed to be uniformly distributed throughout all organs and tissues of the body. A classification of isotopes of the element for the purposes of bone dosimetry is, therefore, not required.

### References

1. *ICRP Publication 23, Report of the ICRP Task Group on Reference Man.* Pergamon Press, Oxford, 1975.
2. Volwiler, W., Goldsworthy, P. D., MacMartin, M. P., Wood, P. A., Mackay, I. R. and Fremont-Smith, K. (1955). Biosynthetic Determination with Radioactive Sulphur of Turn-over Rates of Various Plasma Proteins in Normal and Cirrhotic Man. *J. Clin. Invest.* **34**, 1126–1146.
3. Minski, M. and Vennart, J. (1971). Maximum Permissible Intakes of $^{35}$S-L-Methionine and $^{35}$S-Sodium Sulphate Deduced from Experiments on Rats. *Health Phys.* **20**, 1–9.
4. Dziewiatkowski, D. D. (1949). On the Utilization of Exogenous Sulphate Sulphur by the Rat in the Formation of Ethereal Sulphates as indicated by the use of Sodium Sulphate Labelled with Radioactive Sulphur. *J. Biol. Chem.*, **178**, 389–393.
5. Dziewiatkowski, D. D. Sulfur. In: *Mineral Metabolism*, Vol. 2, Part B, Eds. Comar, C. L. and Bronner, F. Academic Press, New York, 1962
6. ICRP Task Group on Lung Dynamics (1966). Deposition and retention models for internal dosimetry of the human respiratory tract. *Health Phys.* **12**, 173–207.
7. Vennart, J. and Ash, P. J. N. D. (1976). Derived Limits for $^{35}$S in Food and Air. *Health Phys.* **30**, 291–294.
8. Walser, M., Seldin, D. W. and Grollman, A. (1953). An Evaluation of Radiosulphate for the Determination of the Volume of Extracellular Fluid in Man and Dogs. *J. Clin. Invest.* **32**, 299–311.
9. Andrews, J. R., Swarm, R. L., Schlachter, L., Brace, K. C., Rubin, P., Bergenstal, D. M., Gump, H., Siefel, S. and Swain, R. W. (1960). The Effects of One Curie of Sulphur 35 Administered Intravenously as Sulphate to a Man with Advanced Chondrosarcoma. *Am. J. Roent.* **83**, 123–134.
10. Maass, A. R., Flanagan, T. L., Blackburn, D. and Smith, M. (1963). Accidental Personnel Exposure to Elemental S[35]. *Health Phys.* **9**, 731–740.
11. Gottschalk, R. G., Alpert, L. K. and Miller, P. O. (1959). The Use of Large Amounts of Radioactive Sulfur in Patients with Advanced Chondrosarcomas II Distribution and Tissue Irradiation. *Cancer Res.* **19**, 1078–1085.
12. Woodard, H. Q., Pentlow, K. S., Mayer, K., Laughlin, J. S. and Marcove, R. C. (1976). Distribution and Retention of $^{35}$S-Sodium Sulphate in Man. *J. Nucl. Med.* **17**, 285–289.

Annual limits on intake, ALI(Bq) and derived air concentrations, DAC(Bq/m³)
(40 h/wk) for isotopes of sulphur

| Radionuclide | | Oral | | Inhalation | |
|---|---|---|---|---|---|
| | | | | Class D | Class W |
| | | $f_1 = 8 \times 10$ | $f_1 = 1 \times 10^{-1}$ | $f_1 = 8 \times 10^{-1}$ | $f_1 = 8 \times 10^{-1}$ |
| $^{35}$S | ALI | $4 \times 10^8$ | $2 \times 10^8$ $(3 \times 10^8)$ LLI Wall | $6 \times 10^8$ | $8 \times 10^7$ |
| | DAC | | | $3 \times 10^5$ | $3 \times 10^4$ |

| | Vapours | |
|---|---|---|
| | | Inhalation |
| $^{35}$S | ALI | $5 \times 10^8$ |
| | DAC | $2 \times 10^5$ |

# METABOLIC DATA FOR CHLORINE

## 1. Metabolism

Data from Reference Man[1]

| | |
|---|---|
| Chlorine content of the body | 95 g |
| of soft tissues | 81 g |
| Daily intake in food and fluids | 5.2 g |

## 2. Metabolic Model

### (a) *Uptake to blood*

Balance studies on 6 young men[2] demonstrated that the fractional absorption of dietary chloride from the gastrointestinal tract is greater than 0.9. Wiseman[3] has concluded that the fractional absorption of the daily intake of chloride by the small intestine is greater than 0.8. In this report $f_1$ is taken to be 1 for all compounds of chlorine.

### (b) *Inhalation classes*

The ICRP Task Group on Lung Dynamics[4] assigns chlorides of all elements either to inhalation class D or to inhalation class W. For information concerning the classification of a chloride of a particular element the metabolic data for that element, or the Task Group Report, should be consulted.

| Inhalation Class | $f_1$ |
|:---:|:---:|
| D | 1 |
| W | 1 |
| Y | — |

### (c) *Distribution and retention*

Two healthy humans injected intravenously with $^{36}Cl$-labelled sodium chloride[5] eliminated half the $^{36}Cl$ in about 6 days when fed on a diet high in chloride. However, when placed on a diet low in chloride the half-life of retention was in the region of 30 days. These half-lives can be compared with the 10 day biological half-life found for the retention of labelled sodium bromide in humans, since bromide and chloride metabolism have been shown to be very similar in man.[6] In this report chloride leaving the transfer compartment is assumed to distribute uniformly among all organs and tissues of the body[1] and to be retained in those organs and tissues with a biological half-life of 10 days.

### (d) *Behaviour of daughters*

$^{39}Cl$ decays to $^{39}Ar$ which, because of its long radioactive half-life, is assumed to escape from the body without decaying.

## 3. Classification of Isotopes for Bone Dosimetry

Because chlorine is assumed to be uniformly distributed throughout the body a classification of isotopes of the element for the purpose of bone dosimetry is not required.

# References

1. *ICRP Publication 23, Report of the ICRP Task Group on Reference Man.* Pergamon Press, Oxford, 1975.
2. Burrill, M. W., Freeman, S. and Ivy, A. C. (1945). Sodium Potassium and Chloride Excretion of Human Subjects Exposed to a Simulated Altitude of Eighteen Thousand Feet. *J. Biol. Chem.* **157**, 297–302.
3. Wiseman, G. Absorption from the Intestine. Academic Press, London, 1964, pp. 201–204.
4. ICRP Task Group on Lung Dynamics (1966). Deposition and retention models for internal dosimetry of the human respiratory tract. *Health Phys.* **12**, 173–207.
5. Ray, C. T., Burch, G. E. and Threefoot, S. A. (1952). Biologic Decay Rates of Chloride in Normal and Diseased Man Determined with Long-life Radiochlorine, $Cl^{36}$. *J. Lab. Clin. Med.* **39**, 673–696.
6. Reid, A. F., Forbes, G. B., Bondurant, J. and Etheridge, J. (1956). Estimation of Total Body Chlorine in Man by Radio-Bromide Dilution. *J. Lab. Clin. Med.* **48**, 63–68.

Annual limits on intake, ALI(Bq) and derived air concentrations DAC(Bq/m³)
(40 h/wk) for isotopes of chlorine

| Radionuclide | | Oral | Inhalation | |
|---|---|---|---|---|
| | | | Class D | Class W |
| | | $f_1 = 1$ | $f_1 = 1$ | $f_1 = 1$ |
| $^{36}Cl$ | ALI | $6 \times 10^7$ | $9 \times 10^7$ | $9 \times 10^6$ |
| | DAC | — | $4 \times 10^4$ | $4 \times 10^3$ |
| $^{38}Cl$ | ALI | $6 \times 10^8$ ($9 \times 10^8$) ST Wall | $2 \times 10^9$ | $2 \times 10^9$ |
| | DAC | — | $6 \times 10^5$ | $7 \times 10^5$ |
| $^{39}Cl$ | ALI | $8 \times 10^8$ ($1 \times 10^9$) ST Wall | $2 \times 10^9$ | $2 \times 10^9$ |
| | DAC | — | $8 \times 10^5$ | $9 \times 10^5$ |

# METABOLIC DATA FOR ARGON

No metabolic model is proposed for argon. As explained in Chapter 8 of Part 1 of this report, for those radionuclides that emit either photons or $\beta$-particles of considerable energy, exposure in a cloud of a radioactive noble gas is usually limited by external irradiation, since dose-equivalent rates from gas absorbed in tissue or contained in the lungs will be negligible in comparison with the dose-equivalent rates to tissues from external irradiation. An exception is $^{37}$Ar, which emits very low energy Auger electrons and X-rays (see Preface). The recommended DACs for $^{39}$Ar and $^{41}$Ar are therefore based on considerations of external irradiation and for $^{37}$Ar on dose equivalent in lung.

Derived air concentrations, DAC(Bq/m$^3$)
(40 h/wk) for isotopes of argon

| Radionuclide | Inhalation |
|---|---|
| $^{37}$Ar | $5 \times 10^{10}$ |

| Radionuclide | Semi-infinite cloud | 1000 m$^3$ room | 500 m$^3$ room | 100 m$^3$ room |
|---|---|---|---|---|
| $^{39}$Ar | $7 \times 10^6$ | $7 \times 10^6$ | $7 \times 10^6$ | $7 \times 10^6$ |
| | $(5 \times 10^8)$ | $(7 \times 10^9)$ | $(9 \times 10^9)$ | $(2 \times 10^{10})$ |
| | Skin | Skin | Skin | Skin |
| $^{41}$Ar | $1 \times 10^5$ | $2 \times 10^6$ | $2 \times 10^6$ | $2 \times 10^6$ |
| | | $(3 \times 10^6)$ | $(3 \times 10^6)$ | $(6 \times 10^6)$ |
| | | Skin | Skin | Skin |

# METABOLIC DATA FOR POTASSIUM

## 1. Metabolism

Data from Reference Man[1]

|  | |
|---|---|
| Potassium content of the body | 140 g |
| of soft tissue | 120 g |
| Daily intake in food and fluids | 3.3 g |

## 2. Metabolic Model

### (a) *Uptake to blood*

Absorption of potassium from the gastrointestinal tract is known to be nearly complete[1] and $f_1$ has therefore been taken to be 1.

### (b) *Inhalation classes*

Following the recommendations of the ICRP Task Group on Lung Dynamics[2] all compounds of potassium are considered to be in inhalation class D.

| Inhalation Classs | $f_1$ |
|---|---|
| D | 1 |
| W | - - |
| Y | — |

### (c) *Distribution and retention*

Potassium is fairly uniformly distributed throughout all the organs and tissues of the body[1] and is readily exchangeable between all these organs and tissues[3,4]. The whole body retention of the element is, therefore, best described by a single exponential. The half-life of this exponential can be estimated from the total body content of stable potassium and its rate of excretion. Using the data given in Reference Man[1] a half-life of 30 days is obtained and this value has been adopted here.

Thus, potassium entering the transfer compartment is assumed to be instantaneously translocated to all organs and tissues of the body where it is retained with a biological half-life of 30 days. It is also assumed that potassium is uniformly distributed throughout the body at all times after its entry into the transfer compartment.

## 3. Classification of Isotopes for the Purpose of Bone Dosimetry

Because potassium is assumed to be uniformly distributed throughout the body, a classification of isotopes of the element for the purpose of bone dosimetry is not required.

# References

1. *ICRP Publication 23, Report of the ICRP Task Group on Reference Man.* Pergamon Press, Oxford, 1975.
2. ICRP Task Group on Lung Dynamics (1966). Deposition and retention models for internal dosimetry of the human respiratory tract. *Health Phys.* **12**, 173–207.
3. Sollman, R. A Manual of Pharmacology. W. B. Saunders, 1975, pp. 1039–1041.
4. Wilde, W. S. Potassium. In: *Mineral Metabolism*, Vol. 2, Part B, Eds. Comar, C. L. and Bronner, F. Academic Press, London, 1962.

Annual limits on intake, ALI(Bq) and derived air concentrations, DAC(Bq/m$^3$) (40 h/wk) for isotopes of potassium

| Radionuclide | | Oral | Inhalation Class D |
|---|---|---|---|
| | | $f_1 = 1$ | $f_1 = 1$ |
| $^{40}$K | ALI | $1 \times 10^7$ | $1 \times 10^7$ |
| | DAC | — | $6 \times 10^3$ |
| $^{42}$K | ALI | $2 \times 10^8$ | $2 \times 10^8$ |
| | DAC | — | $7 \times 10^4$ |
| $^{43}$K | ALI | $2 \times 10^8$ | $3 \times 10^8$ |
| | DAC | — | $1 \times 10^5$ |
| $^{44}$K | ALI | $8 \times 10^8$ ($1 \times 10^9$) ST Wall | $2 \times 10^9$ |
| | DAC | — | $1 \times 10^6$ |
| $^{45}$K | ALI | $1 \times 10^9$ ($2 \times 10^9$) ST Wall | $4 \times 10^9$ |
| | DAC | — | $2 \times 10^6$ |

# METABOLIC DATA FOR CALCIUM

## 1. Metabolism

Data from Reference Man[1]

| | |
|---|---|
| Calcium content of the body | 1000 g |
| of bone | $\sim 1000$ g |
| of soft tissue | 3 g |
| Daily intake in food and fluids | 1.1 g |

## 2. Metabolic Model

(a) *Uptake to blood*

The fractional absorption of dietary calcium from the gastrointestinal tract of humans is usually about 0.3 although variations from 0.12 to 0.7 have been observed.[1] The fractional absorption of calcium chloride following oral administration has been observed to be in the range 0.4 to 0.8 for individuals on a low calcium diet.[2] In this report, $f_1$ is taken to be 0.3 for all compounds of calcium.

(b) *Inhalation classes*

Following the recommendations of the ICRP Task Group on Lung Dynamics[3] all compounds of calcium have been assigned to inhalation class W.

| Inhalation Class | $f_1$ |
|---|---|
| D | — |
| W | 0.3 |
| Y | — |

(c) *Distribution and retention*

A very comprehensive model for the retention of calcium in adults has been developed by the ICRP Task Group on Alkaline Earth Metabolism in Man.[4]

The total numbers of spontaneous nuclear transformations in soft tissue, cortical bone and trabecular bone during the 50 years following the introduction of 1 Bq of a radioactive isotope of calcium into the transfer compartment can be derived from the retention functions given in the Task Group report.[4]

## 3. Classification of Isotopes for Bone Dosimetry

As discussed in Chapter 7, isotopes of the alkaline earths with radioactive half-lives of greater than 15 days are assumed to be uniformly distributed throughout the volume of mineral bone, whereas isotopes with radioactive half-lives of less than 15 days are assumed to be uniformly distributed in a thin layer over bone surfaces. Thus, when deriving values of absorbed fraction (see Table 7.4, Chapter 7, vol. 1) $^{41}$Ca and $^{45}$Ca are assumed to be uniformly distributed throughout the volume of mineral bone at all times following their deposition in the skeleton, whereas $^{47}$Ca is assumed to be uniformly distributed over bone surfaces at all times following its deposition in the skeleton.

## References

1. *ICRP Publication 23, Report of the ICRP Task Group on Reference Man.* Pergamon Press, Oxford, 1975.
2. Samachson, J. (1963). Plasma Values after Oral [45]Calcium and [85]Strontium. As an Index of Absorption. *Clin. Sci.* **25**, 17–26.
3. ICRP Task Group on Lung Dynamics (1966). Deposition and retention models for internal dosimetry of the human respiratory tract. *Health Phys.* **12**, 173–207.
4. *ICRP Publication 20, Task Group Report on Alkaline Earth Metabolism in Adult Man.* Pergamon Press, Oxford, 1973.

Annual limits on intake, ALI(Bq) and derived air concentrations,
DAC(Bq/m$^3$) (40 h/wk) for isotopes of calcium

| Radionuclide | | Oral | Inhalation Class W |
|---|---|---|---|
| | | $f_1 = 3 \times 10^{-1}$ | $f_1 = 3 \times 10^{-1}$ |
| [41]Ca | ALI | $1 \times 10^8$ | $1 \times 10^8$ |
| | | $(1 \times 10^8)$ | $(1 \times 10^8)$ |
| | | Bone Surf | Bone Surf |
| | DAC | — | $6 \times 10^4$ |
| [45]Ca | ALI | $6 \times 10^7$ | $3 \times 10^7$ |
| | DAC | — | $1 \times 10^4$ |
| [47]Ca | ALI | $3 \times 10^7$ | $3 \times 10^7$ |
| | DAC | — | $1 \times 10^4$ |

# METABOLIC DATA FOR CHROMIUM

## 1. Metabolism

Data from Reference Man[1]

| | |
|---|---|
| Chromium content of the body | $<6.6$ mg |
| of soft tissue | 1.8 mg |
| of the skeleton | $<4.8$ mg |
| Daily intake in food and fluids | 0.15 mg |

## 2. Metabolic Model

(a) *Uptake to blood*

Various reviews[1-3] have discussed the gastrointestinal absorption of chromium. The value of $f_1$ can vary from less than $5 \times 10^{-3}$ to 0.1 or more depending upon the compound administered. In this report $f_1$ is taken to be 0.01 for chromium in the trivalent state and 0.1 for chromium in the hexavalent state.

(b) *Inhalation classes*

The ICRP Task Group on Lung Dynamics[4] assigned oxides and hydroxides of chromium to inhalation class Y, halides and nitrates to inhalation class W and all other compounds of the element to inhalation class D. Experiments using dogs[5] indicate that $Cr_2O_3$ and $CrCl_3$ are properly assigned to inhalation classes Y and W respectively. In view of these experimental data the Task Group's classification is adopted in this report. It is also assumed, for the purposes of radiological protection, that chromium entering the gastrointestinal tract following inhalation is in the hexavalent state and that an $f_1$ of 0.1 is appropriate.

| Inhalation Class | $f_1$ |
|---|---|
| D | 0.1 |
| W | 0.1 |
| Y | 0.1 |

(c) *Distribution and retention*

The retention of chromium in the body is very dependent upon the chemical form administered. Chromium in the form of sodium chromate has a marked affinity for erythrocytes whereas chromic chloride does not penetrate the erythrocyte membrane.[6] Chromium in erythrocytes disappears from the circulation of normal persons with a biological half-life of approximately 30 days.[6] However, normal subjects excrete about 25% of chromium intravenously administered in the chromic form in the first 24 h post-injection.[7] In rats[8] the whole body retention of $^{51}Cr$ intravenously injected as $^{51}Cr\ Cl_3.6H_2O$ is well described by a function of the form:

$$R(t) = 0.43e^{-0.693t/0.5} = 0.43\ e^{-} + 0.32e^{-0.693t/5.9} + 0.25e^{-0.693t/83.4}$$

Experiments on rats[9] also show that $^{51}Cr$ intravenously injected as $^{51}CrCl_3$ is preferentially concentrated in the testes and spleen during the first few days post-injection. Data from Reference Man[1] do not show any preferential concentration of stable chromium in these organs, but do indicate that the element is preferentially concentrated in bone.

It is probable that by the time chromium enters the systemic circulation after its ingestion or inhalation it will have been reduced to its chromic form and will not bind appreciably to erythrocytes. Therefore, in this report, a metabolic model appropriate to the chromic $(3+)$ form has been adopted.

Chromium entering the transfer compartment is assumed to be retained there with a biological half-life of 0.5 days. Of chromium leaving the transfer compartment 0.3 is assumed to go directly to excreta and 0.05 to bone, where it is assumed to be retained with a biological half-life of 1 000 days. The remainder of chromium leaving the transfer compartment is assumed to be uniformly distributed throughout all organs and tissues of the body other than the skeleton. The remaining fraction of chromium distributed in this way is 0.65 and of this 0.4 is assumed to be retained with a biological half-life of 6 days while 0.25 is assumed to be retained with a biological half-life of 80 days.

## 3. Classification of Isotopes for Bone Dosimetry

In the absence of any relevant data concerning the distribution of chromium in the skeleton it is assumed in this report that isotopes of chromium with radioactive half-lives of less than 15 days are uniformly distributed on bone surfaces and that isotopes of chromium with radioactive half-lives of greater than 15 days are uniformly distributed throughout the volume of mineral bone.

### References

1. *ICRP Publication 23, Report of the ICRP Task Group on Reference Man.* Pergamon Press, Oxford, 1975.
2. Underwood, E. J. Trace Elements in Human and Animal Nutrition, 3rd Edition. Academic press, London, 1971, pp. 253–266.
3. Mertz, W. (1969). Chromium Occurrence and Function in Biological Systems. *Physiol. Rev.* **49**, 163–239.
4. ICRP Task Group on Lung Dynamics (1966). Deposition and retention models for internal dosimetry of the human respiratory tract. *Health Phys.* **12**, 173–207.
5. Morrow, P. E., Gibb, F. R., Davies, H. and Fisher, M. (1968). Dust Removal from the Lung Parenchyma: an Investigation of Clearance Simulants. *Toxicol. and Applied Pharmacology* **12**, 372–396.
6. Korst, D. R. Blood Volume and Red Cell Survival. In: *Principles of Nuclear Medicine*, Ed. Wagner, H. N. W. B. Saunders, Philadelphia, 1968, pp. 429–471.
7. Doisy, R. J., Streeten, D. H. P., Souma, M. L., Kalafer, M. E., Rekant, S. I. and Dalakos, T. G. Metabolism of [51]Chromium in Human Subjects 1. Normal, Elderly and Diabetic Subjects. In: *Newer Trace Elements in Nutrition*, Eds. Mertz, W. and Cornatzen, W. E. Marcel Dekker, New York, 1971.
8. Mertz, W., Roginski, E. E. and Reba, R. C. (1965). Biological Activity and Fate of Trace Quantities of Intravenous Chromium (III) in the Rat. *Am. J. Physiol.* **209**, 489–494.
9. Hopkins, L. L. (1965). Distribution in the Rat of Physiological Amounts of Injected Cr[51] (III) with Time. *Am. J. Physiol.* **209**, 731–735.

Annual limits on intake, ALI(Bq) and derived air concentrations, DAC(Bq/m³)
(40 h/wk) for isotopes of chromium

| Radionuclide | | Oral | | Inhalation | | |
|---|---|---|---|---|---|---|
| | | | | Class D | Class W | Class Y |
| | | $f_1 = 1 \times 10^{-1}$ | $f_1 = 1 \times 10^{-2}$ | $f_1 = 1 \times 10^{-1}$ | $f_1 = 1 \times 10^{-1}$ | $f_1 = 1 \times 10^{-1}$ |
| [48]Cr | ALI | $2 \times 10^8$ | $2 \times 10^8$ | $4 \times 10^8$ | $3 \times 10^8$ | $3 \times 10^8$ |
| | DAC | — | — | $2 \times 10^5$ | $1 \times 10^5$ | $1 \times 10^5$ |
| [49]Cr | ALI | $1 \times 10^9$ | $1 \times 10^9$ | $3 \times 10^9$ | $4 \times 10^9$ | $3 \times 10^9$ |
| | DAC | — | — | $1 \times 10^6$ | $2 \times 10^6$ | $1 \times 10^6$ |
| [51]Cr | ALI | $1 \times 10^9$ | $1 \times 10^9$ | $2 \times 10^9$ | $9 \times 10^8$ | $7 \times 10^8$ |
| | DAC | — | — | $7 \times 10^5$ | $4 \times 10^5$ | $3 \times 10^5$ |

# METABOLIC DATA FOR IRON

## 1. Metabolism

Data from Reference Man[1]

| | |
|---|---|
| Iron content of the body | 4.2 g |
| of soft tissues | 3.3 g |
| Daily intake in food and fluids | 16 mg |

## 2. Metabolic Model

(a) *Uptake to blood*

Absorption of iron from the gastrointestinal tract has been reviewed by several authors[1-3]. It has been found to be dependent upon a number of factors, the amount of iron in the diet and its chemical form, the body's need for that iron and the presence or absence of interfering substances in the diet. Absorption of ferrous salts has generally been considered to be greater than that of ferric salts[1] but the distinction is not clear cut[2,3]. In this report $f_1$ is taken as 0.1 for all compounds of iron.

(b) *Inhalation classes*

Experiments on rats and dogs[4-6] indicate that both $FeCl_3$ and $Fe_2O_3$ should be assigned to inhalation class W. In man, experiments with $^{51}Cr$-labelled sub-micron particles of ferric oxide indicate a clearance half-life for the $^{51}Cr$ of 270 days.[7] This long half-life is probably associated with $^{51}Cr$ leached from the ferric oxide, since other studies have shown that ferric oxide is cleared from the lungs with a biological half-life of 70 days.[8] In this report oxides, hydroxides and halides of iron are assigned to inhalation class W and all other commonly occurring compounds of the element are assigned to inhalation class D.

| Inhalation Class | $f_1$ |
|---|---|
| D | 0.1 |
| W | 0.1 |
| Y | — |

(c) *Distribution and retention*

The distribution and retention of iron has been the subject of various reviews.[2,3,9,10] In adult man some 70% of the total body iron is bound in haemoglobin and most of the rest is associated with the iron storage compounds ferritin and haemosiderin in the reticuloendothelial system.

Studies covering the first hundred days post-ingestion[2] indicate a component of iron retention with a half-life of 600 days. However, these studies cannot reasonably be used to estimate long-term iron retention since little recycling of labelled haemoglobin iron will have occurred over this period. In this report the half-life of iron retention in the body has been taken to be 2 000 days, in accordance with the metabolic data given for Reference Man[1] and a value of 0.1 for $f_1$.

The distribution of iron among the organs and tissues of the body has been determined from the data given in Reference Man.[1] Thus, of iron leaving the transfer compartment, fractions of 0.08 and 0.013 are assumed to be translocated to liver and spleen respectively. The remaining fraction of iron leaving the transfer compartment is assumed to be uniformly distributed throughout all other organs and tissues of the body. Iron translocated to any organ or tissue is assumed to be retained there with a biological half-life of 2 000 days.

## 3. Classification of Isotopes for Bone Dosimetry

Iron is assumed to be uniformly distributed throughout all organs and tissues of the body other than the liver and spleen. Therefore, a classification of isotopes of the element for the purposes of bone dosimetry is not required.

## References

1. *ICRP Publication 23, Report of the ICRP Task Group on Reference Man.* Pergamon Press, Oxford, 1975.
2. Price, D. C. Iron Turnover in Man. In: *Dynamic Clinical Studies with Radioisotopes.* AEC Symposium Series 3, Eds. Kindeley, R. M. and Tauxe, W. N. (1964), pp. 537–563.
3. Underwood, E. J. Trace Elements in Human and Animal Nutrition 3rd Ed. Academic Press, New York, 1971, pp. 14–56.
4. Fisher, M. V., Morrow, P. E. and Yuile, C. L. (1973). Effects of Freund's Complete Adjuvant Upon Clearance of Iron-59 Oxide from Rat Lungs. *J. Reticuloendothel. Soc.* **13,** 536–556.
5. Morrow, P. E., Gibb, F. R. and Johnson, L. (1964). Clearance of Insoluble Dust from the lower Respiratory Tract. *Health Phys.* **10,** 543–555.
6. Morrow, P. E., Gibb, F. R., Davies, H. and Fisher, M. (1968). Dust removal from the Lung Parenchyma: an Investigation of Clearance Simulants. *Toxicol. and Applied Pharmacology* **12,** 372–396.
7. Ramsden, D., Waite, D. A. Inhalation of Insoluble Iron-Oxide Particles in the Submicron Range. In: *Assessment of Radioactive Contamination in Man.* IAEA, Vienna, 1972, pp. 65–81.
8. Albert, R. E., Lippmann, M., Spiegelman, J., Strehlow, C., Briscoe, W., Wolfson, P. and Nelson, N. The Clearance of Radioactive Particles from the Human Lung. In: *Inhaled Particles and Vapours II,* Ed. Davies, C. N. Pergamon Press, Oxford, 1967, pp. 361–378.
9. Moore, C. V. and Dubach, R. Iron. In: *Mineral Metabolism,* Vol. 2, Part B, Eds. Comar, C. L. and Bronner, F. Academic Press, New York, 1962, pp. 287–348.
10. Finch, C. A., Deubelbeiss, K., Cook, J. D., Eschbach, J. W., Horker, L. A., Funk, D. D., Marsaglia, G., Hillman, R. S., Slichter, S., Adamson, J. W., Canzoni, A. and Giblett, E. R. (1970). Ferrokinetics in Man. *Medicine* **49,** 17–53.

Annual limits on intake, ALI(Bq) and derived air concentrations, DAC(Bq/m³)
(40 h/wk) for isotopes of iron

| | | | Inhalation | |
|---|---|---|---|---|
| Radionuclide | | Oral | Class D | Class W |
| | | $f_1 = 1 \times 10^{-1}$ | $f_1 = 1 \times 10^{-1}$ | $f_1 = 1 \times 10^{-1}$ |
| $^{52}$Fe | ALI | $3 \times 10^7$ | $1 \times 10^8$ | $9 \times 10^7$ |
| | DAC | — | $5 \times 10^4$ | $4 \times 10^4$ |
| $^{55}$Fe | ALI | $3 \times 10^8$ | $7 \times 10^7$ | $2 \times 10^8$ |
| | DAC | — | $3 \times 10^4$ | $6 \times 10^4$ |
| $^{59}$Fe | ALI | $3 \times 10^7$ | $1 \times 10^7$ | $2 \times 10^7$ |
| | DAC | — | $5 \times 10^3$ | $8 \times 10^3$ |
| $^{60}$Fe | ALI | $1 \times 10^6$ | $2 \times 10^5$ | $7 \times 10^5$ |
| | DAC | — | $1 \times 10^2$ | $3 \times 10^2$ |

# METABOLIC DATA FOR COPPER

## 1. Metabolism

Data from Reference Man[1]

|  |  |
|---|---|
| Copper content of the body | 72 mg |
| of soft tissue | 65 mg |
| Daily intake in food and fluids | 3.5 mg |

## 2. Metabolic Model

(a) *Uptake to blood*

The fractional uptake of copper from the gastrointestinal tract of man has been reported to be in the range 0.32 to 0.9.[1-4] In this report $f_1$ is taken to be 0.5 for all compounds of copper.

(b) *Inhalation classes*

The ICRP Task Group on Lung Dynamics[5] assigned oxides and hydroxides of copper to inhalation class Y, sulphides, halides and nitrates to inhalation class W and all other inorganic compounds of the element to inhalation class D. In the absence of any relevant experimental data this classification has been adopted here.

| Inhalation Class | $f_1$ |
|---|---|
| D | 0.5 |
| W | 0.5 |
| Y | 0.5 |

(c) *Distribution and retention*

The metabolism of copper has been reviewed by various authors.[6, 7] There is some variation of the concentration of the stable element in different organs and tissues of the body[1], for example the concentration in brain, liver, salivary glands and pancreas is about five times higher than the concentration in muscle and spleen.[8, 9] A biological half-life for the retention of copper in the body can be estimated from the data given in Reference Man[1] if an $f_1$ of 0.5 is assumed. The estimated biological half-life is 41 days.

In this report it is assumed that of copper leaving the transfer compartment fractions 0.1, 0.1 and 0.006 are translocated to the liver, brain and pancreas respectively. The remaining fraction of copper leaving the transfer compartment is assumed to be uniformly distributed throughout all other organs and tissues of the body. Copper translocated to any organ or tissue of the body is assumed to be retained there with a biological half-life of 40 days.

## 3. Classification of Isotopes for Bone Dosimetry

Copper is assumed to be uniformly distributed throughout all organs and tissues of the body other than the liver, brain and pancreas. A classification of isotopes of the element for the purpose of bone dosimetry is, therefore, not required.

# References

1. *ICRP Publication* 23, *Report of the ICRP Task Group on Reference Man.* Pergamon Press, Oxford, 1975.
2. Wiseman, G. Absorption from the Intestine. Academic Press, London, 1964.
3. Sternlieb, I. (1967). Gastrointestinal Copper Absorption in Man. *Gastroenterology* 52, 1038–1041.
4. Strickland, G. T., Beckner, W. M. and Leu, Mei-Ling (1972). Absorption of Copper in Homozygotes and Heterozygotes for Wilson's Disease and Controls: Isotope Tracer Studies with $^{67}$Cu and $^{64}$Cu. *Clin. Sci.* 43, 617–625.
5. ICRP Task Group on Lung Dynamics (1966). Deposition and retention models for internal dosimetry of the human respiratory tract. *Health Phys.* 12, 173–207.
6. Adelstein, S. J. and Vallee, B. L. Copper. In: *Mineral Metabolism*, Vol. 2, Part B, Eds., Comar, C. L. and Bronner, F. Academic Press, London, 1962, pp. 371–401.
7. Underwood, E. J. Trace Elements in Human and Animal Nutrition. Academic Press, New York, 1971, pp. 57–115.
8. Cartwright, G. E. and Wintrobe, M. M. (1964) Copper Metabolism in Normal Subjects. *Am. J. Clin. Nutr.* 14, 224–232.
9. De Jorge, F. B., Canelas, H. M., Dias, J. C. and Cury, L. (1964). Studies on Copper Metabolism III. Copper Contents of Saliva of Normal Subjects and of Salivary Glands and Pancreas of Autopsy Material. *Clin. Chim. Acta* 9, 148–150.

Annual limits on intake, ALI(Bq) and derived air concentrations DAC(Bq/m$^3$) (40 h/wk) for isotopes of copper

| Radionuclide | | Oral | Inhalation | | |
| --- | --- | --- | --- | --- | --- |
| | | | Class D | Class W | Class Y |
| | | $f_1 = 5 \times 10^{-1}$ | $f_1 = 5 \times 10^{-1}$ | $f_1 = 5 \times 10^{-1}$ | $f_1 = 5 \times 10^{-1}$ |
| $^{60}$Cu | ALI | $1 \times 10^9$ ($1 \times 10^9$) ST Wall | $3 \times 10^9$ | $4 \times 10^9$ | $4 \times 10^9$ |
| | DAC | — | $1 \times 10^6$ | $2 \times 10^6$ | $2 \times 10^6$ |
| $^{61}$Cu | ALI | $5 \times 10^8$ | $1 \times 10^9$ | $2 \times 10^9$ | $1 \times 10^9$ |
| | DAC | — | $5 \times 10^5$ | $6 \times 10^5$ | $5 \times 10^5$ |
| $^{64}$Cu | ALI | $4 \times 10^8$ | $1 \times 10^9$ | $9 \times 10^8$ | $8 \times 10^8$ |
| | DAC | — | $5 \times 10^5$ | $4 \times 10^5$ | $3 \times 10^5$ |
| $^{67}$Cu | ALI | $2 \times 10^8$ | $3 \times 10^8$ | $2 \times 10^8$ | $2 \times 10^8$ |
| | DAC | — | $1 \times 10^5$ | $8 \times 10^4$ | $7 \times 10^4$ |

# METABOLIC DATA FOR ZINC

## 1. Metabolism

Data from Reference Man[1]

| | |
|---|---|
| Zinc content of the body | 2.3 g |
| of soft tissues | 1.8 g |
| Daily intake in food and fluids | 0.013 g |

## 2. Metabolic Model

(a) *Uptake to blood*

The fractional uptake of orally administered zinc from the gastrointestinal tract of adult man has been estimated to be in the range 0.31 to 0.51,[1] although there is some indication that this fraction is dependent on the daily intake of the element[2, 3] and that fractional uptakes of as much as 0.9 may sometimes occur.[4] Uptake by man appears to be dependent on fasting state and uptake by experimental animals varies with dietary zinc level.[5] In this report $f_1$ is taken to be 0.5 for all compounds of the element.

(b) *Inhalation classes*

The ICRP Task Group on Lung Dynamics[6] assigned oxides and hydroxides of zinc to inhalation class Y, halides, phosphates and sulphides to inhalation class W and sulphates to inhalation class D. Experiments on dogs[7] suggest that $Zn(NO_3)_2$ and $Zn_3(PO_4)_2$ should be assigned to inhalation class Y.

In man, very little information on inhaled zinc is available. In one case of accidental exposure $^{65}$Zn was rapidly cleared from the lungs except for a small component which was retained for a period of several months.[8] However, in this case the compound or compounds of the element involved are not known.

In this report all commonly occurring compounds of zinc are assigned to inhalation class Y.

| Inhalation Class | $f_1$ |
|---|---|
| D | — |
| W | — |
| Y | 0.5 |

(c) *Distribution and retention*

The metabolism of zinc has been reviewed by various authors.[2, 3] In man Spencer *et al.*[9] found the whole body biological retention of the element to be well described by the function,

$$R(t) = 0.25e^{-0.693t/12} + 0.75e^{-0.693t/320}$$

Richmond *et al.*[4] have studied the retention of orally-administered zinc in mice, rats, dogs and man. In each of these species there were two components of whole body retention. In the 4 human subjects between 0.15 and 0.26 of the retained $^{65}$Zn was associated with a biological half-life of between 4.5 and 26 days and the remainder was associated with a biological half-life of between 387 and 478 days.

At early times post-administration the highest concentration of zinc is found in the liver, with kidney, spleen and pancreas exhibiting rather lower concentrations of the element.[9,10] However, from the data of Richmond et al.[4] it is clear that zinc is more tenaciously retained in the skeleton than it is in other tissues and this is in agreement with the distribution of stable zinc given for Reference Man.[1]

In this report it is assumed that of zinc leaving the transfer compartment, 0.2 is translocated to the skeleton and retained there with a biological half-life of 400 days. The remainder of zinc entering the transfer compartment is assumed to be uniformly distributed throughout all other organs and tissues of the body. Of zinc translocated to these other organs and tissues, fractions 0.3 and 0.7 are assumed to be retained with biological half-lives of 20 and 400 days respectively.

### 3. Classification of Isotopes for Bone Dosimetry

There appear to be no relevant data available on the microscopic distribution of zinc in the skeleton. In this report it is assumed that $^{65}Zn$, with a radioactive half-life of 245 days, is uniformly distributed throughout the volume of mineral bone at all times following its deposition in the skeleton and that all the other, shorter lived, radioactive isotopes of zinc are uniformly distributed over bone surfaces at all times following their deposition in the skeleton.

### References

1. ICRP Publication 23, Report of the ICRP Task Group on Reference Man. Pergamon Press, Oxford, 1975.
2. Vallee, B. L. Zinc. In: Mineral Metabolism, Vol. 2, Part B, Eds. Comar, C. L. and Bronner, F. Academic Press, New York, 1962, pp. 443–482.
3. Underwood, E. J. Trace Elements in Human and Animal Nutrition. Academic Press, New York, 1971, pp. 217–221.
4. Richmond, C. R., Furchner, J. E., Trafton, G. A. and Langham, W. H. (1962). Comparative Metabolism of Radionuclides in Mammals—I Uptake and Retention of Orally Administered Zn$^{65}$ by Four Mammalian Species. Health Phys. 8, 481–489.
5. Furchner, J. E. and Richmond, C. R. (1962). Effect of Dietary Zinc on the Absorption of Orally Administered Zn$^{65}$. Health Phys. 8, 35–40.
6. ICRP Task Group on Lung Dynamics (1966). Deposition and retention models for internal dosimetry of the human respiratory tract. Health Phys. 12, 173–207.
7. Morrow, P. E., Gibb, F. R., Davies, H. and Fisher, M. (1968). Dust Removal from the Lung Parenchyma: an Investigation of Clearance Simulants. Toxicol. and Applied Pharmacology 12, 372–376.
8. Newton, D. and Holmes, A. (1966). A Case of Accidental Inhalation of Zinc-65 and Silver-110m. Rad. Res. 29, 403–412.
9. Spencer, H., Rosoff, B., Feldstein, A., Cohn, S. and Gusmano, E. (1965). Metabolism of Zinc-65 in Man. Rad. Res. 24, 432–445.
10. Siegel, E., Craig, F. A., Crystal, M. M. and Siegel, E. P. (1961). Distribution of $^{65}Zn$ in the Prostrate and other Organs of Man. Brit. J. Cancer 15, 647–664.

Annual limits on intake, ALI(Bq) and derived air concentrations, DAC(Bq/m³) (40 h/wk) for isotopes of zinc

| Radionuclide | | Oral | Inhalation Class Y |
|---|---|---|---|
| | | $f_1 = 5 \times 10^{-1}$ | $f_1 = 5 \times 10^{-1}$ |
| $^{62}Zn$ | ALI | $5 \times 10^7$ | $1 \times 10^8$ |
| | DAC | — | $4 \times 10^4$ |
| $^{63}Zn$ | ALI | $9 \times 10^8$ $(9 \times 10^8)$ ST Wall | $3 \times 10^9$ |
| | DAC | — | $1 \times 10^6$ |
| $^{65}Zn$ | ALI | $1 \times 10^7$ | $1 \times 10^7$ |
| | DAC | — | $4 \times 10^3$ |
| $^{69m}Zn$ | ALI | $2 \times 10^8$ | $3 \times 10^8$ |
| | DAC | — | $1 \times 10^5$ |
| $^{69}Zn$ | ALI | $2 \times 10^9$ | $5 \times 10^9$ |
| | DAC | — | $2 \times 10^6$ |
| $^{71m}Zn$ | ALI | $2 \times 10^8$ | $6 \times 10^8$ |
| | DAC | — | $3 \times 10^5$ |
| $^{72}Zn$ | ALI | $4 \times 10^7$ | $4 \times 10^7$ |
| | DAC | — | $2 \times 10^4$ |

# METABOLIC DATA FOR BROMINE

## 1. Metabolism

Data from Reference Man[1]

| | |
|---|---|
| Bromine content of the body | 200 mg |
| of soft tissue | 170 mg |
| Daily intake in food and fluids | 7.5 mg |

## 2. Metabolic Model

(a) *Uptake to blood*

Nearly all orally administered bromide appears in the urine[2] and by analogy with chlorine, $f_1$ is taken to be 1 for all compounds of this element.

(b) *Inhalation classes*

The ICRP Task Group on Lung Dynamics[3] assigns bromides of all elements either to inhalation class D or to inhalation class W. For information concerning the classification of a bromide of a particular element the metabolic data for that element, or the Task Group Report, should be consulted.

| Inhalation Class | $f_1$ |
|:---:|:---:|
| D | 1 |
| W | 1 |
| Y | — |

(c) *Distribution and retention*

The metabolic behaviour of bromide in the human body is known to be very similar to that of chloride.[4] The metabolic model used for chlorine is, therefore, also used for bromine. Bromide leaving the transfer compartment is assumed to be uniformly distributed among all organs and tissues of the body where it is retained with a biological half-life of 10 days.

(d) *Behaviour of daughters*

$^{83}$Br decays to $^{83m}$Kr which has a radioactive half-life of 114 min. The elimination of krypton from the body has been studied in various animals and in man.[5, 6] Components of elimination with half-lives of a few minutes are observed but, so also are components with half-lives of several hours. After reviewing these data it has been assumed that 0.2 of the $^{83m}$Kr produced decays at its site of origin and that the other 0.8 escapes from the body without decaying.

## 3. Classification of Isotopes for Bone Dosimetry

Because bromine is assumed to be uniformly distributed throughout the body, a classification of isotopes of the element for the purpose of bone dosimetry is not required.

# References

1. *ICRP Publication* 23, *Report of the ICRP Task Group on Reference Man*. Pergamon Press, Oxford, 1975.
2. Söremark, R. (1960). Excretion of Bromide Ions by Human Urine. *Acta Physiol. Scand.* **50**, 306–310.
3. ICRP Task Group on Lung Dynamics (1966). Deposition and retention models for internal dosimetry of the human respiratory tract. *Health Phys.* **12**, 173–207.
4. Reid, A. F., Forbes, C. B., Bondurant, J. and Etheridge, J. (1956). Estimation of Total Body Chlorine in Man by Radio-Bromide Dilution. *J. Lab. Clin. Med.* **48**, 63–68.
5. Kirk, W. P. and Morken, D. A. (1975). *In vivo* Kinetic Behaviour and Whole-Body Partition Coefficients for [85]Kr in Guinea Pigs. *Health Phys.* **28**, 263–273.
6. Hytten, F. E., Taylor, K. and Taggart, N. (1966). Measurement of Total Body Fat in Man by Absorption of [85]Kr. *Clin. Sci.* **31**, 111–119.

Annual limits on intake, ALI(Bq) and derived air concentrations, DAC(Bq/m$^3$)
(40 h/wk) for isotopes of bromine

| Radionuclide | | Oral | Inhalation | |
|---|---|---|---|---|
| | | | Class D | Class W |
| | | $f_1 = 1$ | $f_1 = 1$ | $f_1 = 1$ |
| $^{74m}$Br | ALI | $5 \times 10^8$ ($8 \times 10^8$) ST Wall | $1 \times 10^9$ | $2 \times 10^9$ |
| | DAC | — | $6 \times 10^5$ | $6 \times 10^5$ |
| $^{74}$Br | ALI | $8 \times 10^8$ ($1 \times 10^9$) ST Wall | $3 \times 10^9$ | $3 \times 10^9$ |
| | DAC | — | $1 \times 10^6$ | $1 \times 10^6$ |
| $^{75}$Br | ALI | $1 \times 10^9$ ($1 \times 10^9$) ST Wall | $2 \times 10^9$ | $2 \times 10^9$ |
| | DAC | — | $7 \times 10^5$ | $8 \times 10^5$ |
| $^{76}$Br | ALI | $1 \times 10^8$ | $2 \times 10^8$ | $2 \times 10^8$ |
| | DAC | — | $7 \times 10^4$ | $7 \times 10^4$ |
| $^{77}$Br | ALI | $6 \times 10^8$ | $9 \times 10^8$ | $7 \times 10^8$ |
| | DAC | — | $4 \times 10^5$ | $3 \times 10^5$ |
| $^{80m}$Br | ALI | $8 \times 10^8$ | $6 \times 10^8$ | $5 \times 10^8$ |
| | DAC | — | $3 \times 10^5$ | $2 \times 10^5$ |
| $^{80}$Br | ALI | $2 \times 10^9$ ($3 \times 10^9$) ST Wall | $7 \times 10^9$ | $8 \times 10^9$ |
| | DAC | — | $3 \times 10^6$ | $3 \times 10^6$ |
| $^{82}$Br | ALI | $1 \times 10^8$ | $2 \times 10^8$ | $1 \times 10^8$ |
| | DAC | — | $6 \times 10^4$ | $6 \times 10^4$ |
| $^{83}$Br | ALI | $2 \times 10^9$ ($3 \times 10^9$) ST Wall | $2 \times 10^9$ | $2 \times 10^9$ |
| | DAC | — | $1 \times 10^6$ | $1 \times 10^6$ |
| $^{84}$Br | ALI | $7 \times 10^8$ ($1 \times 10^9$) ST Wall | $2 \times 10^9$ | $2 \times 10^9$ |
| | DAC | — | $9 \times 10^5$ | $1 \times 10^6$ |

# METABOLIC DATA FOR RUBIDIUM

## 1. Metabolism

Data from Reference Man[1]

| | |
|---|---|
| Rubidium content of the body | 0.68 g |
| of soft tissue | 0.47 g |
| Daily intake in food and fluids | 2.2 mg |

## 2. Metabolic Model

(a) *Uptake to blood*

Experiments on man indicate that rubidium is almost completely absorbed from the gastrointestinal tract.[2] In this report $f_1$ is taken as 1 for all compounds of rubidium.

(b) *Inhalation classes*

The ICRP Task Group on Lung Dynamics[3] assigned compounds of rubidium to inhalation class D and, in the absence of any relevant experimental data, this classification is adopted here.

| Inhalation Class | $f_1$ |
|---|---|
| D | 1 |
| W | — |
| Y | — |

(c) *Distribution and retention*

Rubidium like potassium is approximately uniformly distributed among all organs and tissues of the body except for mineral bone which has about three times the average whole body concentration.[1] The whole body retention of rubidium can be reasonably well described by a single exponential[2, 4—7] although a small component of rapid clearance has been observed.[2, 5] This single exponential has a biological half-life of between 32 and 57 days in normal subjects, the mean value being 44 days.[2]

In this report it is assumed that of rubidium leaving the transfer compartment a fraction, 0.25, is translocated to the skeleton. The remaining fraction of rubidium leaving the transfer compartment is assumed to be uniformly distributed throughout all other organs and tissues of the body. Rubidium translocated to any organ or tissue of the body, including the skeleton, is assumed to be retained there with a biological half-life of 44 days.

(d) *Behaviour of daughters*

$^{79}$Rb decays to $^{79}$Kr which has a radioactive half-life of 35 h and $^{81}$Rb to $^{81}$Kr which has a radioactive half-life of $2.1 \times 10^5$ years. For purposes of radiological protection it is assumed that $^{79}$Kr and $^{81}$Kr escape from the body without decaying. $^{83}$Rb decays to $^{83m}$Kr which has a radioactive half-life of 114 min. It is assumed that 0.2 of the $^{83m}$Kr produced decays at its site of origin and that the other 0.8 escapes from the body without decaying. (See metabolic data for bromine p. 24.)

## 3. Classification of Isotopes for Bone Dosimetry

It is known that bone tissue fluid has a much higher concentration of potassium than does plasma.[8] Therefore, because of the chemical similarities between rubidium and potassium, it is reasonable to suppose that the higher than average concentration of rubidium in the skeleton is due to a high concentration of the element in bone tissue fluid. For this reason it is assumed that radioactive isotopes of rubidium are uniformly distributed throughout mineral bone at all times following their deposition in the skeleton.

### References

1. *ICRP Publication 23, Report of the ICRP Task Group on Reference Man.* Pergamon Press, Oxford, 1975.
2. Lloyd, R. D., Mays, C. W., McFarland, S. S., Zundel, W. S. and Tyler, F. H. (1973). Metabolism of $^{83}$Rb and $^{137}$Cs in persons with Muscle Disease. *Rad. Res.* **54**, 463–478.
3. ICRP Task Group on Lung Dynamics (1966). Deposition and retention models for internal dosimetry of the human respiratory tract. *Health Phys.* **12**, 173–207.
4. Wood, O. L. (1969). Comparison of Naturally Occurring Rubidium and Potassium in Human Erythrocytes, Plasma and Urine. *Health Phys.* **17**, 513–514.
5. Iinuma, T., Watari, K., Nagai, T., Iwashima, K. and Yamagata, N. (1967). Comparative Studies of $^{132}$Cs and $^{86}$Rb turnover in Man using a Double-tracer Method. *J. Rad. Res.* **8-3-4**, 100–115.
6. Threefoot, S. A., Ray, C. T. and Burch, G. E. (1955). Study of the use of Rb$^{86}$ as a Tracer for the measurement of Rb$^{86}$ and K$^{39}$ Space and Mass in Intact Man with and without Congestive Heart Failure. *J. Lab. Clin. Med.* **45**, 408–430.
7. Richmond, C. R. (1958). Retention and Excretion of Radionuclides of the Alkali Metals by Five Mammalian Species. *USAEC Report LA*-2207, p. 139.
8. Triffit, J. T., Terepka, A. R. and Neuman, W. F. (1968). A comparative Study of the Exchange in vivo of Major Constituents of Bone Mineral. *Calc. Tiss. Res.* **2**, 165–176.

Annual limits on intake, ALI(Bq) and derived air concentrations, DAC(Bq/m$^3$) (40 h/wk) for isotopes of rubidium

| Radionuclide | | Oral | Inhalation Class D |
|---|---|---|---|
| | | $f_1 = 1$ | $f_1 = 1$ |
| $^{79}$Rb | ALI | $1 \times 10^9$ $(2 \times 10^9)$ ST Wall | $4 \times 10^9$ |
| | DAC | — | $2 \times 10^6$ |
| $^{81m}$Rb | ALI | $9 \times 10^9$ $(1 \times 10^{10})$ ST Wall | $1 \times 10^{10}$ |
| | DAC | — | $5 \times 10^6$ |
| $^{81}$Rb | ALI | $1 \times 10^9$ | $2 \times 10^9$ |
| | DAC | — | $8 \times 10^5$ |
| $^{82m}$Rb | ALI | $4 \times 10^8$ | $7 \times 10^8$ |
| | DAC | — | $3 \times 10^5$ |
| $^{83}$Rb | ALI | $2 \times 10^7$ | $4 \times 10^7$ |
| | DAC | — | $2 \times 10^4$ |
| $^{84}$Rb | ALI | $2 \times 10^7$ | $3 \times 10^7$ |
| | DAC | — | $1 \times 10^4$ |
| $^{86}$Rb | ALI | $2 \times 10^7$ | $3 \times 10^7$ |
| | DAC | — | $1 \times 10^4$ |
| $^{87}$Rb | ALI | $4 \times 10^7$ | $6 \times 10^7$ |
| | DAC | — | $2 \times 10^4$ |
| $^{88}$Rb | ALI | $7 \times 10^8$ $(1 \times 10^9)$ ST Wall | $2 \times 10^9$ |
| | DAC | — | $1 \times 10^6$ |
| $^{89}$Rb | ALI | $1 \times 10^9$ $(2 \times 10^9)$ ST Wall | $5 \times 10^9$ |
| | DAC | — | $2 \times 10^6$ |

# METABOLIC DATA FOR YTTRIUM

## 1. Metabolism

The total body content and normal daily intake of yttrium are not given in Reference Man.[1] However, the total liver content of yttrium is given as 1.6 mg and the total trabecular bone content as less than 4.5 mg.

## 2. Metabolic Model

### (a) *Uptake to blood*

In this report $f_1$ is taken as $10^{-4}$ for all compounds of yttrium, since studies on rats[2,3], and dogs[4] have demonstrated that there is little uptake of this element from the gastrointestinal tract.

### (b) *Inhalation classes*

The ICRP Task Group on Lung Dynamics[5] assigned oxides and hydroxides of yttrium to inhalation class Y and all other compounds of the element to inhalation class W. This classification is supported by experiments on dogs[6] which indicate that $YCl_3$ behaves as a class W material.

| Inhalation class | $f_1$ |
|:---:|:---:|
| D | — |
| W | $10^{-4}$ |
| Y | $10^{-4}$ |

### (c) *Distribution and retention*

Experiments on rats[7] indicate that some 30% of injected yttrium is deposited in the liver and 10% is distributed among all other soft tissues. This is in broad agreement with the distribution of $^{91}Y$ in dogs[6] following inhalation of $^{91}YCl_3$ and also with studies on the distribution and retention of $^{91}Y$ in rabbits to 30 days post-injection.[8] The data on retention in rabbits[8] indicate that yttrium is tenaciously retained in all organs and tissues of the body.

In this report it is assumed that of yttrium leaving the transfer compartment 0.25 goes directly to excreta, 0.5 is translocated to the skeleton, 0.15 is translocated to the liver and 0.1 is uniformly distributed throughout all other organs and tissues of the body. It is also assumed that yttrium not going from the transfer compartment directly to excretion is retained indefinitely in the body. This latter assumption is appropriate for the purposes of radiological protection since none of the isotopes of yttrium considered in this report has a radioactive half-life of greater than 105 days.

## 3. Classification of Isotopes for Bone Dosimetry

Since the chemistry of yttrium resembles that of the actinides and since none of the isotopes of yttrium considered in this report has a radioactive half-life of greater than 110 days, radioisotopes of yttrium are assumed to be uniformly distributed over bone surfaces at all times following their deposition in the skeleton.

## References

1. *ICRP Publication 23, Report of the ICRP Task Group on Reference Man.* Pergamon Press, Oxford, 1975.
2. Hamilton, J. G., Scott, K., Chaikoff, I. C., Fishler, M. C., Entermann, C., Overstreet, R., Jacobson, L., Kaplan, M. and Greenberg, D. M. (1943). Metabolism of Fission Products. MDDC-1143.
3. Marcus, C. S. and Lengemann, F. W. (1962). Use of Radioyttrium to Study Food Movement in the Small Intestine of the Rat. *J. Nutrition* **76**, 179–182.
4. Nold, N. M., Hayes, R. L. and Comar, C. L. (1960). Internal Radiation Dose Measurements in Live Experimental Animals—II. *Health Phys.* **4**, 86–100.
5. Report of the ICRP Task Group on Lung Dynamics (1966). Deposition and retention models for internal dosimetry of the human respiratory tract. *Health Phys.* **12**, 173–207.
6. McClellan, R. O. and Rupprecht, F. S. *Fission Product Inhalation Program Annual Report 1966–1967*, pp. 40–64. Lovelace Foundation Report (LF-38).
7. Durbin, P. W. (1960). Metabolic Characteristics within a Chemical Family. *Health Phys.* **2**, 225–238.
8. Lloyd, E. (1961). The Relative Distributions of Radioactive Yttrium and Strontium and the Secondary Deposition of $^{90}Y$ built up from $^{90}Sr$. *Int. J. Rad. Biol.* **3**, 475–492.

Annual limits on intake, ALI(Bq) and derived air concentrations, DAC(Bq/m$^3$)
(40 h/wk) for isotopes of yttrium

| Radionuclide | | Oral | Inhalation | |
|---|---|---|---|---|
| | | | Class W | Class Y |
| | | $f_1 = 1 \times 10^{-4}$ | $f_1 = 1 \times 10^{-4}$ | $f_1 = 1 \times 10^{-4}$ |
| $^{86m}Y$ | ALI | $8 \times 10^8$ | $2 \times 10^9$ | $2 \times 10^9$ |
| | DAC | — | $9 \times 10^5$ | $8 \times 10^5$ |
| $^{86}Y$ | ALI | $5 \times 10^7$ | $1 \times 10^8$ | $1 \times 10^8$ |
| | DAC | — | $5 \times 10^4$ | $5 \times 10^4$ |
| $^{87}Y$ | ALI | $8 \times 10^7$ | $1 \times 10^8$ | $1 \times 10^8$ |
| | DAC | — | $5 \times 10^4$ | $5 \times 10^4$ |
| $^{88}Y$ | ALI | $4 \times 10^7$ | $9 \times 10^6$ | $9 \times 10^6$ |
| | DAC | — | $4 \times 10^3$ | $4 \times 10^3$ |
| $^{90m}Y$ | ALI | $3 \times 10^8$ | $5 \times 10^8$ | $4 \times 10^8$ |
| | DAC | — | $2 \times 10^5$ | $2 \times 10^5$ |
| $^{90}Y$ | ALI | $2 \times 10^7$ $(2 \times 10^7)$ LLI Wall | $3 \times 10^7$ | $2 \times 10^7$ |
| | DAC | — | $1 \times 10^4$ | $9 \times 10^3$ |
| $^{91m}Y$ | ALI | $5 \times 10^9$ | $9 \times 10^9$ | $6 \times 10^9$ |
| | DAC | — | $4 \times 10^6$ | $2 \times 10^6$ |
| $^{91}Y$ | ALI | $2 \times 10^7$ $(2 \times 10^7)$ LLI Wall | $6 \times 10^6$ | $4 \times 10^6$ |
| | DAC | — | $3 \times 10^3$ | $2 \times 10^3$ |
| $^{92}Y$ | ALI | $1 \times 10^8$ | $3 \times 10^8$ | $3 \times 10^8$ |
| | DAC | — | $1 \times 10^5$ | $1 \times 10^5$ |
| $^{93}Y$ | ALI | $4 \times 10^7$ | $1 \times 10^8$ | $9 \times 10^7$ |
| | DAC | — | $4 \times 10^4$ | $4 \times 10^4$ |
| $^{94}Y$ | ALI | $8 \times 10^8$ $(1 \times 10^9)$ ST Wall | $3 \times 10^9$ | $3 \times 10^9$ |
| | DAC | — | $1 \times 10^6$ | $1 \times 10^6$ |
| $^{95}Y$ | ALI | $1 \times 10^9$ $(2 \times 10^9)$ ST Wall | $6 \times 10^9$ | $5 \times 10^9$ |
| | DAC | — | $2 \times 10^6$ | $2 \times 10^6$ |

# METABOLIC DATA FOR TECHNETIUM

## 1. Metabolism

No data are given in Reference Man[1] for technetium.

## 2. Metabolic Model

### (a) Uptake to blood

Beasley et al.[2] reported that the fractional absorption of technetium, in the form of pertechnetate, from the gastrointestinal tract of man is about 0.95. However, Hays[3] has shown that orally administered pertechnetate is erratically absorbed with marked variability in the timing and extent of absorption. In rats the fractional absorption of technetium chloride from the gastrointestinal tract was found to be about 0.5.[4,5] In this report $f_1$ is taken to be 0.8 for all compounds of the element. However, it should be noted that this value is certainly unduly large for particular radiopharmaceuticals e.g. $^{99m}$Tc-sulphur colloid.[6]

### (b) Inhalation classes

The ICRP Task Group on Lung Dynamics[7] assigned oxides, hydroxides, halides and nitrate of technetium to inhalation class W and all other compounds of the element to inhalation class D. Experiments with $^{99m}$Tc-pertechnetate in man[8] are in agreement with this classification and it is adopted here.

| Inhalation Class | $f_1$ |
|---|---|
| D | 0.8 |
| W | 0.8 |
| Y | — |

### (c) Distribution and retention

In man technetium, administered intravenously as pertechnetate, is concentrated in the thyroid, gastrointestinal tract and liver[2, 9−11].

The whole body retention of pertechnetate in man following intravenous injection[2] is well fitted by a function of the form;

$$R(t) = 0.76e^{-0.693t/1.6} + 0.19e^{-0.693t/3.7} + 0.043e^{-0.693t/22}$$

However, to describe the distribution of pertechnetate in the body at early times after intravenous infusion a complex multi-compartmental model is required.[11]

In this report a simple model for the distribution and retention of pertechnetate has been adopted. This model is considered appropriate for the purposes of radiological protection.

Of technetium leaving the transfer compartment 0.04 is assumed to be translocated to the thyroid where it is retained with a biological half-life of 0.5 days.[10] Further fractions 0.1 and 0.03 are assumed to be translocated to the stomach wall and liver respectively. The remaining fraction of technetium leaving the transfer compartment is assumed to be uniformly distributed throughout all organs and tissues of the body other than the thyroid, stomach wall and liver. Of technetium translocated to any organ or tissue of the body other than the thyroid fractions 0.75,

0.20 and 0.05 are assumed to be retained with biological half-lives of 1.6, 3.7 and 22 days respectively. For technetium the biological half-life in the transfer compartment is taken to be 0.02 day.

(d) *Radiopharmaceuticals*

It is emphasized that the metabolic model set out above is, in general, inappropriate for $^{99m}$Tc-labelled radiopharmaceuticals other than pertechnetates. For guidance on radiological protection when working with such compounds the reader is referred to *ICRP Publications 17* and *25* (Refs. 12 and 13).

## 3. Classification of Isotopes for Bone Dosimetry

Technetium is assumed to be uniformly distributed throughout all organs and tissues of the body other than the thyroid, stomach wall and liver. Therefore, a classification of isotopes of the element for the purpose of bone dosimetry is not required.

## References

1. *ICRP Publication 23, Report of the ICRP Task Group on Reference Man*. Pergamon Press, Oxford, 1975.
2. Beasley, T. M., Palmer, H. E. and Nelp, W. B. (1966). Deposition and Excretion of Technetium in Humans. *Health Phys.*, **12**, 1425–1435.
3. Hays, M. T. (1973). $^{99m}$Tc-Pertechnetate transport in man: absorption after subcutaneous and oral administration: secretion into saliva and gastric juice. *J. Nucl. Med.* **14**, 331–335.
4. Hamilton, J. G. *Medical and Health Physics Division, Quarterly Report*. University of California, UCRL-98 (1948) p. 8.
5. Sullivan, M. F., Graham, T. M., Cataldo, D. A. and Schreckhise, R. G. Absorption and retention of inorganic and originally incorporated technetium-95 by rats and guinea pigs. In: *Pacific Northwest Laboratory Annual Report for 1977*, Part 1 Biomedical Sciences, Feb. 1978, PNL-2500PT1.
6. Van Kirk, O., Chafetz, N., Cooke, S., Taylor, A. and Larson, S. M. (1978). Imaging of the bowel with technetium—an aid in gallium studies. *J. Nucl. Med.* **19**, 69–70.
7. ICRP Task Group on Lung Dynamics (1966). Deposition and retention models for internal dosimetry of the human respiratory tract. *Health Phys.* **12**, 173–207.
8. Cooke, D. J. and Lander, H. (1971). Inhalation Pulmonary Scintiphotography using pertechnetate. *Am. J. Roent.* **113**, 682–689.
9. Harper, P. V., Lathrop, K. A., McCordle, R. J. and Andros, G. The use of technetium-99m as a clinical scanning agent for thyroid, liver and brain. In: *Medical Radioisotope Scanning*, Vol. 2 (IAEA, Vienna, 1964) pp. 33–45.
10. McAfee, J. G., Fueger, C. F., Stern, H. S., Wagner, H. N. Jr. and Migata, T. (1964). $^{99m}$Tc pertechnetate for brain scanning. *J. Nucl. Med.* **5**, 811–827.
11. Hays, M. T. and Berman, M. (1977). Pertechnetate distribution in man after intravenous infusion: a compartmental model. *J. Nucl. Med.* **18**, 898–904.
12. *ICRP Publication 17, Protection of the patient in radionuclide investigations*. Pergamon Press, Oxford, 1971.
13. *ICRP Publication 25, Handling, storage, use and disposal of unsealed radionuclides in hospitals and medical research establishments*. Pergamon Press, Oxford, 1977.

Annual limits on intake, ALI(Bq) and derived air concentrations, DAC(Bq/m$^3$)
(40 h/wk) for isotopes of technetium

| Radionuclide | | Oral | Inhalation Class D | Inhalation Class W |
|---|---|---|---|---|
| | | $f_1 = 8 \times 10^{-1}$ | $f_1 = 8 \times 10^{-1}$ | $f_1 = 8 \times 10^{-1}$ |
| $^{93m}$Tc | ALI | $3 \times 10^9$ | $6 \times 10^9$ | $1 \times 10^{10}$ |
| | DAC | — | $2 \times 10^6$ | $5 \times 10^6$ |
| $^{93}$Tc | ALI | $1 \times 10^9$ | $3 \times 10^9$ | $4 \times 10^9$ |
| | DAC | — | $1 \times 10^6$ | $2 \times 10^6$ |
| $^{94m}$Tc | ALI | $7 \times 10^8$ | $2 \times 10^9$ | $2 \times 10^9$ |
| | DAC | — | $7 \times 10^5$ | $9 \times 10^5$ |
| $^{94}$Tc | ALI | $3 \times 10^8$ | $7 \times 10^8$ | $9 \times 10^8$ |
| | DAC | — | $3 \times 10^5$ | $4 \times 10^5$ |
| $^{96m}$Tc | ALI | $6 \times 10^9$ | $1 \times 10^{10}$ | $9 \times 10^9$ |
| | DAC | — | $4 \times 10^6$ | $4 \times 10^6$ |
| $^{96}$Tc | ALI | $7 \times 10^7$ | $1 \times 10^8$ | $8 \times 10^7$ |
| | DAC | — | $5 \times 10^4$ | $3 \times 10^4$ |
| $^{97m}$Tc | ALI | $2 \times 10^8$ | $2 \times 10^8$ $(3 \times 10^8)$ ST Wall | $4 \times 10^7$ |
| | DAC | — | $1 \times 10^5$ | $2 \times 10^4$ |
| $^{97}$Tc | ALI | $1 \times 10^9$ | $2 \times 10^9$ | $2 \times 10^8$ |
| | DAC | — | $8 \times 10^5$ | $9 \times 10^4$ |
| $^{98}$Tc | ALI | $4 \times 10^7$ | $6 \times 10^7$ | $1 \times 10^7$ |
| | DAC | — | $2 \times 10^4$ | $5 \times 10^3$ |
| $^{99m}$Tc | ALI | $3 \times 10^9$ | $6 \times 10^9$ | $9 \times 10^9$ |
| | DAC | — | $2 \times 10^6$ | $4 \times 10^6$ |
| $^{99}$Tc | ALI | $1 \times 10^8$ | $2 \times 10^8$ $(2 \times 10^8)$ ST Wall | $2 \times 10^7$ |
| | DAC | — | $8 \times 10^4$ | $1 \times 10^4$ |
| $^{101}$Tc | ALI | $3 \times 10^9$ $(5 \times 10^9)$ ST Wall | $1 \times 10^{10}$ | $1 \times 10^{10}$ |
| | DAC | — | $5 \times 10^6$ | $6 \times 10^6$ |
| $^{104}$Tc | ALI | $8 \times 10^8$ $(1 \times 10^9)$ ST Wall | $3 \times 10^9$ | $3 \times 10^9$ |
| | DAC | — | $1 \times 10^6$ | $1 \times 10^6$ |

# METABOLIC DATA FOR RUTHENIUM

## 1. Metabolism

No data are given in Reference Man[1] for Ruthenium.

## 2. Metabolic Model

(a) *Uptake to blood*

Furchner *et al.*[2] have estimated the fractional absorption of $^{106}RuCl_3$ from the gastrointestinal tract to be 0.035, a value in good agreement with results reported by Thompson[3], Burykina[4] and Bruce.[5] Although nitrosyl-ruthenium complexes are rather better absorbed than the chloride[5] it is considered sufficient, for the purposes of radiological protection, to take $f_1$ to be 0.05 for all commonly occurring compounds of the element.

(b) *Inhalation classes*

Experiments with beagles[6] have demonstrated that $^{106}RuO_2$ is avidly retained in the lungs with a biological half-life of about 2 000 days. This is in agreement with the conclusions of the ICRP Task Group on Lung Dynamics[7] who classified oxides and hydroxides of ruthenium as inhalation class Y, halides as inhalation class W and all other compounds as inhalation class D. Their recommendations are therefore adopted in this report.

| Inhalation Class | $f_1$ |
|:---:|:---:|
| D | 0.05 |
| W | 0.05 |
| Y | 0.05 |

(c) *Distribution and retention*

Furchner *et al.*[2] have determined the biological retention of ruthenium in various species following both oral and intravenous administration. Correcting their data for the radioactive decay of $^{106}Ru$ a whole body retention function of the form;

$$R(t) = 0.15e^{-0.693t/0.3} + 0.35e^{-0.693t/8.0} + 0.30e^{-0.693t/35} + 0.20e^{-0.693t/1000}$$

has been derived as being appropriate to man.

The experiments of Furchner *et al.*[2] demonstrate that in rats at early times post-injection the kidneys contain the highest concentration of ruthenium but that at times greater than 86 days post-injection similar concentrations of ruthenium exist in all organs and tissues of the body.

In this report ruthenium leaving the transfer compartment is assumed to be retained there with a biological half-life of 0.3 days. Of this ruthenium a fraction, 0.15, is assumed to go directly to excreta and the remainder is assumed to become uniformly distributed throughout all organs and tissues of the body. Of this 0.85 of ruthenium translocated to all organs and tissues of the body from the transfer compartment, 0.35 is assumed to be retained with a biological half-life of 8 days, 0.3 is assumed to be retained with a biological half-life of 35 days and 0.2 is assumed to be retained with a biological half-life of 1 000 days.

## 3. Classification of Isotopes for Bone Dosimetry

Because ruthenium is assumed to be uniformly distributed throughout the body a classification of isotopes of the element for the purposes of bone dosimetry is not required.

## References

1. *ICRP Publication 23, Report of the ICRP Task Group on Reference Man.* Pergamon Press, Oxford, 1975.
2. Furchner, J. E., Richmond, C. R. and Drake, G. A. (1971). Comparative Metabolism of Radionuclides in Mammals—VII. Retention of $^{106}$Ru in the Mouse, Rat, Monkey and Dog. *Health Phys.* **21**, 355–365.
3. Thompson, R. C., Weeks, M. H., Hollis, L., Ballou, J. F. and Oakely, W. D. (1958). Metabolism of Radio-Ruthenium in the Rat. *Am. J. Roentg.* **79**, 1026–1044.
4. Burykina, L. N. The Metabolism of Radioactive Ruthenium in the Organism of Experimental Animals. In: *The Toxicology of Radioactive Substances*, Vol. 1. Eds. Letavet, A. A. and Kurlyandskaya, E. B. Pergamon Press, Oxford, 1962, pp. 60–76.
5. Bruce, R. S. and Carr, T. E. F. (1961). Studies in the Metabolism of Carrier-Free Radioruthenium I. *Reactor Sc. Technol., J. nucl. Energy* (Parts A and B), **14**, 9–17.
6. Stuart, B. O. (1970). Long-term Retention and Translocation of Inhaled $^{106}$Ru–$^{106}$RhO$_2$ in Beagles. *Pacific Northwest Lab. Report*, BNWL-1050, pp. 3–43.
7. ICRP Task Group on Lung Dynamics (1966). Deposition and retention models for internal dosimetry of the human respiratory tract. *Health Phys.* **12**, 173–207.

Annual limits on intake, ALI(Bq) and derived air concentrations, DAC(Bq/m$^3$) (40 h/wk)
for isotopes of ruthenium

| Radionuclide | | Oral | Inhalation | | |
|---|---|---|---|---|---|
| | | | Class D | Class W | Class Y |
| | | $f_1 = 5 \times 10^{-2}$ | $f_1 = 5 \times 10^{-2}$ | $f_1 = 5 \times 10^{-2}$ | $f_1 = 5 \times 10^{-2}$ |
| $^{94}$Ru | ALI | $6 \times 10^8$ | $2 \times 10^9$ | $2 \times 10^9$ | $2 \times 10^9$ |
| | DAC | — | $7 \times 10^5$ | $1 \times 10^6$ | $9 \times 10^5$ |
| $^{97}$Ru | ALI | $3 \times 10^8$ | $7 \times 10^8$ | $5 \times 10^8$ | $4 \times 10^8$ |
| | DAC | — | $3 \times 10^5$ | $2 \times 10^5$ | $2 \times 10^5$ |
| $^{103}$Ru | ALI | $7 \times 10^7$ | $6 \times 10^7$ | $4 \times 10^7$ | $2 \times 10^7$ |
| | DAC | — | $3 \times 10^4$ | $2 \times 10^4$ | $1 \times 10^4$ |
| $^{105}$Ru | ALI | $2 \times 10^8$ | $5 \times 10^8$ | $5 \times 10^8$ | $4 \times 10^8$ |
| | DAC | — | $2 \times 10^5$ | $2 \times 10^5$ | $2 \times 10^5$ |
| $^{106}$Ru | ALI | $7 \times 10^6$ $(9 \times 10^6)$ LLI Wall | $3 \times 10^6$ | $2 \times 10^6$ | $4 \times 10^5$ |
| | DAC | — | $1 \times 10^3$ | $8 \times 10^2$ | $2 \times 10^2$ |

# METABOLIC DATA FOR RHODIUM

## 1. Metabolism

No data are given in Reference Man[1] for rhodium.

## 2. Metabolic Model

(a) *Uptake to blood*

There appears to be no information concerning the uptake of rhodium from the gastrointestinal tract. Chemically the element resembles ruthenium[2] and $f_1$ is, therefore, taken to be 0.05 for all its compounds.

(b) *Inhalation classes*

The ICRP Task Group on Lung Dynamics[3] assigned oxides and hydroxides of rhodium to inhalation class Y, halides to inhalation class W and all other compounds of the element to inhalation class D. In the absence of any relevant experimental information this classification has been adopted here.

| Inhalation Class | $f_1$ |
|---|---|
| D | 0.05 |
| W | 0.05 |
| Y | 0.05 |

(c) *Distribution and retention*

There appears to be no information available concerning the distribution and retention of rhodium in any mammalian species. Since the element has similar chemical properties to ruthenium the metabolic model for that element is also adopted for rhodium.

Rhodium entering the transfer compartment is assumed to be retained there with a biological half-life of 0.3 days. Of this rhodium a fraction, 0.15, is assumed to go directly to excretion and the remainder is assumed to become uniformly distributed throughout all organs and tissues of the body. Of this 0.85 of rhodium translocated to all organs and tissues of the body from the transfer compartment 0.35 is assumed to be retained with a biological half-life of 8 days, 0.3 is assumed to be retained with a biological half-life of 35 days and 0.2 is assumed to be retained with a biological half-life of 1 000 days.

## 3. Classification of Isotopes for Bone Dosimetry

Because rhodium is assumed to be uniformly distributed throughout the body a classification of isotopes of the element for the purpose of bone dosimetry is not required.

## References

1. *ICRP Publication 23, Task Group Report on Reference Man*. Pergamon Press, Oxford, 1975.
2. Partington, J. R. *General and Inorganic Chemistry*. Macmillan, London, 1954, pp. 834–835.
3. ICRP Task Group on Lung Dynamics (1966). Deposition and retention models for internal dosimetry of the human respiratory tract. *Health Phys.* **12**, 173–207.

Annual limits on intake, ALI(Bq) and derived air concentrations, DAC(Bq/m$^3$)
(40 h/wk) for isotopes of rhodium

| Radionuclide | | Oral | Inhalation | | |
|---|---|---|---|---|---|
| | | | Class D | Class W | Class Y |
| | | $f_1 = 5 \times 10^{-2}$ | $f_1 = 5 \times 10^{-2}$ | $f_1 = 5 \times 10^{-2}$ | $f_1 = 5 \times 10^{-2}$ |
| $^{99m}$Rh | ALI | $7 \times 10^8$ | $2 \times 10^9$ | $3 \times 10^9$ | $2 \times 10^9$ |
| | DAC | — | $9 \times 10^5$ | $1 \times 10^6$ | $1 \times 10^6$ |
| $^{99}$Rh | ALI | $9 \times 10^7$ | $1 \times 10^8$ | $8 \times 10^7$ | $7 \times 10^7$ |
| | DAC | — | $5 \times 10^4$ | $3 \times 10^4$ | $3 \times 10^4$ |
| $^{100}$Rh | ALI | $6 \times 10^7$ | $2 \times 10^8$ | $1 \times 10^8$ | $1 \times 10^8$ |
| | DAC | — | $8 \times 10^4$ | $6 \times 10^4$ | $6 \times 10^4$ |
| $^{101m}$Rh | ALI | $2 \times 10^8$ | $4 \times 10^8$ | $3 \times 10^8$ | $3 \times 10^8$ |
| | DAC | — | $2 \times 10^5$ | $1 \times 10^5$ | $1 \times 10^5$ |
| $^{101}$Rh | ALI | $8 \times 10^7$ | $2 \times 10^7$ | $3 \times 10^7$ | $6 \times 10^6$ |
| | DAC | — | $8 \times 10^3$ | $1 \times 10^4$ | $2 \times 10^3$ |
| $^{102m}$Rh | ALI | $5 \times 10^7$ $(5 \times 10^7)$ LLI Wall | $2 \times 10^7$ | $1 \times 10^7$ | $4 \times 10^6$ |
| | DAC | — | $8 \times 10^3$ | $6 \times 10^3$ | $2 \times 10^3$ |
| $^{102}$Rh | ALI | $2 \times 10^7$ | $3 \times 10^6$ | $7 \times 10^6$ | $2 \times 10^6$ |
| | DAC | — | $1 \times 10^3$ | $3 \times 10^3$ | $9 \times 10^2$ |
| $^{103m}$Rh | ALI | $2 \times 10^{10}$ | $4 \times 10^{10}$ | $5 \times 10^{10}$ | $4 \times 10^{10}$ |
| | DAC | — | $2 \times 10^7$ | $2 \times 10^7$ | $2 \times 10^7$ |
| $^{105}$Rh | ALI | $1 \times 10^8$ $(1 \times 10^8)$ LLI Wall | $4 \times 10^8$ | $2 \times 10^8$ | $2 \times 10^8$ |
| | DAC | — | $2 \times 10^5$ | $1 \times 10^5$ | $9 \times 10^4$ |
| $^{106m}$Rh | ALI | $3 \times 10^8$ | $9 \times 10^8$ | $1 \times 10^9$ | $1 \times 10^9$ |
| | DAC | — | $4 \times 10^5$ | $6 \times 10^5$ | $5 \times 10^5$ |
| $^{107}$Rh | ALI | $3 \times 10^9$ $(3 \times 10^9)$ ST Wall | $9 \times 10^9$ | $1 \times 10^{10}$ | $9 \times 10^9$ |
| | DAC | — | $4 \times 10^6$ | $4 \times 10^6$ | $4 \times 10^6$ |

# METABOLIC DATA FOR SILVER

## 1. Metabolism

Data from Reference Man[1]

| | |
|---|---|
| Silver content of soft tissues | 790 $\mu$g |
| Daily intake in food and fluids | 70 $\mu$g |

## 2. Metabolic Model

(a) *Uptake to blood*

Experiments on rats, monkeys and dogs indicated that the fractional absorption of silver from the gastrointestinal tract is less than 0.1 if the silver is administered as the nitrate. In this report $f_1$ is taken as 0.05 for all compounds of silver, a value based on the data of Furchner *et al.*[2] for absorption of the nitrate in monkeys and dogs.

(b) *Inhalation classes*

The ICRP Task Group on Lung Dynamics[3] assigned oxides and hydroxides of silver to inhalation class Y, nitrates and sulphides to inhalation class W and all other compounds of the element to inhalation class D. In the absence of any relevant experimental data this classification has been adopted here.

In dogs which inhaled a fume of metallic silver it was found that fractions 0.59, 0.39 and 0.02 of the silver deposited in the lung were retained with biological half-lives of 1.7, 8.4 and 40 days respectively.[4] Therefore, in this report, metallic silver is assigned to inhalation class D.

| Inhalation Class | $f_1$ |
|:---:|:---:|
| D | 0.05 |
| W | 0.05 |
| Y | 0.05 |

(c) *Distribution and retention*

Following intravenous injection of radioactive silver into man external counting demonstrates components of silver retention with half-lives of about 3.5 and 48 days.[5] Similar components of retention were found for an individual who accidentally inhaled $^{110m}$Ag.

Experiments on animals[2] indicate that there is considerable inter-species variability in silver retention and it is, therefore, difficult to extrapolate the available animal data to man.

In the rat, silver is preferentially concentrated in the spleen and brain,[2] whereas in the dog, after inhalation of metallic silver, the liver is the organ to which the bulk of the silver is translocated.

In man the post-mortem data given by Polachek *et al.*[5] indicate that about 40% of systemic silver is found in the liver. The data presented by Newton and Holmes[6] also indicate that the liver is the predominant site of deposition for silver translocated from the lungs following inhalation.

In this report it is assumed that of silver leaving the transfer compartment 0.8 is translocated to the liver and 0.2 is distributed uniformly throughout all other organs and tissues of the body. Of silver going to liver or any other tissue fractions 0.1 and 0.9 are assumed to be retained with biological half-lives of 3.5 and 50 days respectively.[5]

## 3. Classification of Isotopes for Bone Dosimetry

Systemic silver is assumed to be uniformly distributed throughout all organs and tissues of the body other than the liver. Therefore, a classification of isotopes of the element for the purposes of bone dosimetry is not required.

## References

1. *ICRP Publication 23, Report of the ICRP Task Group on Reference Man.* Pergamon Press, Oxford, 1975.
2. Furchner, J. E., Richmond, C. R. and Drake, G. A. (1968). Comparative Metabolism of Radionuclides in Mammals—IV. Retention of Silver-110m in the Mouse, Rat, Monkey and Dog. *Health Phys.* **15**, 505–514.
3. ICRP Task Group on Lung Dynamics (1966). Deposition and retention models for internal dosimetry of the human respiratory tract. *Health Phys.* **12**, 173–207.
4. Phalen, R. F. and Morrow, P. E. (1973). Experimental Inhalation of Metallic Silver. *Health Phys.* **24**, 509–518.
5. Polachek, A. A., Cope, C. B., Willard, R. F. and Enns, T. (1960). Metabolism of Radioactive Silver in A Patient with Carcinoid. *J. Lab. Clin. Med.* **56**, 499–505.
6. Newton, D. and Holmes, A. (1966). A Case of Accidental Inhalation of Zinc-65 and Silver-110m. *Rad. Res.* **29**, 403–412.

Annual limits on intake, ALI(Bq) and derived air concentrations, DAC(Bq/m$^3$)
(40 h/wk) for isotopes of silver

| Radionuclide | | Oral | Inhalation | | |
|---|---|---|---|---|---|
| | | | Class D | Class W | Class Y |
| | | $f_1 = 5 \times 10^{-2}$ | $f_1 = 5 \times 10^{-2}$ | $f_1 = 5 \times 10^{-2}$ | $f_1 = 5 \times 10^{-2}$ |
| $^{102}$Ag | ALI | $2 \times 10^9$ $(2 \times 10^9)$ ST Wall | $7 \times 10^9$ | $8 \times 10^9$ | $7 \times 10^9$ |
| | DAC | — | $3 \times 10^6$ | $3 \times 10^6$ | $3 \times 10^6$ |
| $^{103}$Ag | ALI | $1 \times 10^9$ | $4 \times 10^9$ | $5 \times 10^9$ | $4 \times 10^9$ |
| | DAC | — | $2 \times 10^6$ | $2 \times 10^6$ | $2 \times 10^6$ |
| $^{104m}$Ag | ALI | $1 \times 10^9$ | $4 \times 10^9$ | $5 \times 10^9$ | $4 \times 10^9$ |
| | DAC | — | $1 \times 10^6$ | $2 \times 10^6$ | $2 \times 10^6$ |
| $^{104}$Ag | ALI | $8 \times 10^8$ | $3 \times 10^9$ | $5 \times 10^9$ | $6 \times 10^9$ |
| | DAC | — | $1 \times 10^6$ | $2 \times 10^6$ | $2 \times 10^6$ |
| $^{105}$Ag | ALI | $1 \times 10^8$ | $4 \times 10^7$ | $6 \times 10^7$ | $6 \times 10^7$ |
| | DAC | — | $2 \times 10^4$ | $3 \times 10^4$ | $3 \times 10^4$ |
| $^{106m}$Ag | ALI | $3 \times 10^7$ | $3 \times 10^7$ | $3 \times 10^7$ | $3 \times 10^7$ |
| | DAC | — | $1 \times 10^4$ | $1 \times 10^4$ | $1 \times 10^4$ |
| $^{106}$Ag | ALI | $2 \times 10^9$ $(2 \times 10^9)$ ST Wall | $7 \times 10^9$ | $8 \times 10^9$ | $7 \times 10^9$ |
| | DAC | — | $3 \times 10^6$ | $3 \times 10^6$ | $3 \times 10^6$ |
| $^{108m}$Ag | ALI | $2 \times 10^7$ | $7 \times 10^6$ | $9 \times 10^6$ | $9 \times 10^5$ |
| | DAC | — | $3 \times 10^3$ | $4 \times 10^3$ | $4 \times 10^2$ |
| $^{110m}$Ag | ALI | $2 \times 10^7$ | $5 \times 10^6$ | $7 \times 10^6$ | $3 \times 10^6$ |
| | DAC | — | $2 \times 10^3$ | $3 \times 10^3$ | $1 \times 10^3$ |
| $^{111}$Ag | ALI | $3 \times 10^7$ $(4 \times 10^7)$ LLI Wall | $6 \times 10^7$ $(6 \times 10^7)$ Liver | $3 \times 10^7$ | $3 \times 10^7$ |
| | DAC | — | $2 \times 10^4$ | $1 \times 10^4$ | $1 \times 10^4$ |
| $^{112}$Ag | ALI | $1 \times 10^8$ | $3 \times 10^8$ | $4 \times 10^8$ | $3 \times 10^8$ |
| | DAC | — | $1 \times 10^5$ | $2 \times 10^5$ | $1 \times 10^5$ |
| $^{115}$Ag | ALI | $1 \times 10^9$ $(1 \times 10^9)$ ST Wall | $3 \times 10^9$ | $3 \times 10^9$ | $3 \times 10^9$ |
| | DAC | — | $1 \times 10^6$ | $1 \times 10^6$ | $1 \times 10^6$ |

# METABOLIC DATA FOR CADMIUM

## 1. Metabolism

Data from Reference Man[1]

| | |
|---|---|
| Cadium content of the body | 50 mg |
| of soft tissues | 38 mg |
| Daily intake in food and fluids | 0.15 mg |

## 2. Metabolic Model

(a) *Uptake to blood*

The fractional absorption of dietary cadmium from the gastrointestinal tract is typically about 0.1.[2] Reports of a much larger fraction are probably in error.[2] Several studies[2-7] in various mammalian species have indicated that the fractional absorption of inorganic cadmium from the gastrointestinal tract is usually less than 0.1 and may sometimes be less than 0.01. In man data are available for cadmium in a calf liver suspension,[8] 80% of the metal being protein bound. In this case fractional absorption from the gastrointestinal tract can be estimated to have been in the range 0.06 to 0.25. In this report $f_1$ is taken to be 0.05 for all inorganic compounds of cadmium.

(b) *Inhalation classes*

The ICRP Task Group on Lung Dynamics[9] assigned oxides and hydroxides of cadmium to inhalation class Y, sulphides, halides and nitrates to inhalation class W and all other compounds of the element to inhalation class D. This classification is adopted here, although it is noted that in dogs exposed to near lethal doses of cadmium chloride by inhalation a long term component of lung retention was observed.[10]

| Inhalation Class | $f_1$ |
|---|---|
| D | 0.05 |
| W | 0.05 |
| Y | 0.05 |

(c) *Distribution and retention*

The distribution and retention of cadmium in various mammalian species has been reviewed in detail by Friberg *et al.*[11] Experiments in which radioactive cadmium was given to male volunteers indicate that the biological half-life for cadmium retention in the whole body is not less than 130 days and could be much greater.[8] Comparisons of excretion and total body burden of stable cadmium for various groups of people indicate biological half-lives for cadmium retention in the whole body of between 13 and 47 years.[11]

Stable cadmium is found to be concentrated in the liver and the kidney.[1,11,12] Data given in Reference Man[1] suggest that the fractions of the whole body content of stable cadmium associated with the liver and kidneys are 0.08 and 0.2 respectively. However, data presented by

Friberg *et al.*[11] suggest that the liver and kidneys each contain about 0.3 of the body's total cadmium.

In this report it is assumed that of cadmium leaving the transfer compartment a fraction, 0.3, is translocated to the liver and 0.3 to the kidneys. The remaining fraction of cadmium leaving the transfer compartment is assumed to be uniformly distributed throughout all other organs and tissues of the body. Cadmium translocated to any organ or tissue of the body, including the liver and kidneys, is assumed to be retained with a biological half-life of 25 years.

### 3. Classification of Isotopes for Bone Dosimetry

Cadmium is assumed to be uniformly distributed throughout all organs and tissues of the body other than the kidneys and liver. Therefore, a classification of isotopes of the element for the purposes of bone dosimetry is not required.

### References

1. *ICRP Publication 23, Report of the ICRP Task Group on Reference Man.* Pergamon Press, Oxford, 1975.
2. Nordberg, G. F., Friberg, L. and Piscator, M. In: *Cadmium in the Environment.* CRC Press, Cleveland, 1971, p. 30 and p. 44. Also see *Cadmium in the Environment 2nd Edition,* Friberg, L., Piscator, M., Nordberg, G. F. and Kjellström, T., CRC Press, Cleveland. 1974, p. 27.
3. Decker, C. F., Byerrum, R. U. and Hoppert, C. A. (1975). A Study of the Distribution and Retention of Cadmium in the Albino Rat. *Arch. Biochem.* **66**, 140–145.
4. Cotzias, G. C., Borg, D. C. and Selleck, B. (1961). Virtual Absence of Turnover in Cadmium Metabolism: $Cd^{109}$ Studies in the Mouse. *Am. J. Physiol.* **201**, 927–930.
5. Richmond, C. R., Findlay, J. S. and London, J. E. Whole-body Retention of Cadmium-109 by Mice Following Oral, Intraperitoneal and Intravenous Administration. *Health Division Annual Report, July 1965 to June 1966.* Los Alamos Scientific Laboratory Report, LA-3610-MS, 1966, p. 95.
6. Miller, W. J., Blackman, D. M. and Martin, W. G. (1968). Cadmium Absorption, Excretion and Tissue Distribution following Single Tracer Oral and Intravenous Doses in Young Goats. *J. Dairy Sci.* **51**, 1836–1839.
7. Silva, A. J., Fleshman, D. G. and Shore, B. (1970). The Effects of Sodium Alginate on the Absorption and Retention of Several Divalent Cations. *Health Phys.* **19**, 245–251.
8. Rahola, T., Aaran, R., Miettinen, J. K. (1972). Half-time Studies of Mercury and Cadmium by Whole-Body Counting. In: *Assessment of Radioactive Contamination in Man.* IAEA-SM-150/13 pp. 553–562.
9. ICRP Task Group on Lung Dynamics (1966). Deposition and retention models for internal dosimetry of the human respiratory tract. *Health Phys.* **12**, 173–207.
10. Harrison, H. E., Bunting, N. and Albrink, W. S. (1947). The Effects and Treatment of Inhalation of Cadmium Chloride in the Dog. *J. Ind. Hyg. Toxicol.* **29**, 302–314.
11. Friberg, L., Piscator, M., Nordberg, G. F. and Kjellström, T. *Cadmium in the Environment.* CRC Press, Cleveland, 1974, pp. 23–91.
12. Underwood, E. J. *Trace Elements in Humans and Animal Nutrition.* Academic Press, London, 1971, pp. 267–280.

Annual limits on intake, ALI(Bq) and derived air concentrations, DAC(Bq/m³) (40 h/wk) for isotopes of cadmium

| Radionuclide | | Oral | Inhalation | | |
|---|---|---|---|---|---|
| | | | Class D | Class W | Class Y |
| | | $f_1 = 5 \times 10^{-2}$ | $f_1 = 5 \times 10^{-2}$ | $f_1 = 5 \times 10^{-2}$ | $f_1 = 5 \times 10^{-2}$ |
| $^{104}$Cd | ALI | $8 \times 10^8$ | $2 \times 10^9$ | $4 \times 10^9$ | $4 \times 10^9$ |
| | DAC | — | $1 \times 10^6$ | $2 \times 10^6$ | $2 \times 10^6$ |
| $^{107}$Cd | ALI | $8 \times 10^8$ | $2 \times 10^9$ | $2 \times 10^9$ | $2 \times 10^9$ |
| | DAC | — | $8 \times 10^5$ | $9 \times 10^5$ | $8 \times 10^5$ |
| $^{109}$Cd | ALI | $1 \times 10^7$ $(2 \times 10^7)$ Kidneys | $1 \times 10^6$ $(2 \times 10^6)$ Kidneys | $4 \times 10^6$ $(5 \times 10^6)$ Kidneys | $4 \times 10^6$ |
| | DAC | — | $5 \times 10^2$ | $2 \times 10^3$ | $2 \times 10^3$ |
| $^{113m}$Cd | ALI | $9 \times 10^5$ $(1 \times 10^6)$ Kidneys | $9 \times 10^4$ $(1 \times 10^5)$ Kidneys | $3 \times 10^5$ $(4 \times 10^5)$ Kidneys | $5 \times 10^5$ |
| | DAC | — | $4 \times 10^1$ | $1 \times 10^2$ | $2 \times 10^2$ |
| $^{113}$Cd | ALI | $8 \times 10^5$ $(1 \times 10^6)$ Kidneys | $8 \times 10^4$ $(1 \times 10^5)$ Kidneys | $3 \times 10^5$ $(4 \times 10^5)$ Kidneys | $5 \times 10^5$ |
| | DAC | — | $3 \times 10^1$ | $1 \times 10^2$ | $2 \times 10^2$ |
| $^{115m}$Cd | ALI | $1 \times 10^7$ | $2 \times 10^6$ $(3 \times 10^6)$ Kidneys | $5 \times 10^6$ | $5 \times 10^6$ |
| | DAC | — | $8 \times 10^2$ | $2 \times 10^3$ | $2 \times 10^3$ |
| $^{115}$Cd | ALI | $3 \times 10^7$ $(4 \times 10^7)$ LLI Wall | $5 \times 10^7$ | $5 \times 10^7$ | $5 \times 10^7$ |
| | DAC | — | $2 \times 10^4$ | $2 \times 10^4$ | $2 \times 10^4$ |
| $^{117m}$Cd | ALI | $2 \times 10^8$ | $5 \times 10^8$ | $6 \times 10^8$ | $5 \times 10^8$ |
| | DAC | — | $2 \times 10^5$ | $3 \times 10^5$ | $2 \times 10^5$ |
| $^{117}$Cd | ALI | $2 \times 10^8$ | $4 \times 10^8$ | $6 \times 10^8$ | $5 \times 10^8$ |
| | DAC | — | $2 \times 10^5$ | $3 \times 10^5$ | $2 \times 10^5$ |

# METABOLIC DATA FOR INDIUM

## 1. Metabolism

No data are given in Reference Man[1] for indium.

## 2. Metabolic Model

### (a) *Uptake to blood*

Experiments on rats[2] indicate a fractional absorption from the gastrointestinal tract of about 0.02 for indium trichloride. Toxicity studies[3] have shown that the toxicity of orally administered indium is much less than the toxicity of indium administered intravenously. In view of these results $f_1$ is taken to be 0.02 for all compounds of indium.

### (b) *Inhalation classes*

The ICRP Task Group on Lung Dynamics[4] assigned oxides, hydroxides, halides and nitrates of indium to inhalation class W and all other compounds of the element to inhalation class D. This classification is supported by experiments on the inhalation of indium sesquioxide by rats[5] and is adopted here.

| Inhalation Class | $f_1$ |
|---|---|
| D | 0.02 |
| W | 0.02 |
| Y | — |

### (c) *Distribution and retention*

In the mouse[3,6] and rat[2] the highest concentrations of indium in the first few days post-injection are found in kidney, bone, liver and spleen. However, experiments with rabbits[7] indicate that indium is concentrated in bone marrow rather than in mineral bone.

In this report it is assumed that of indium leaving the transfer compartment fractions 0.3, 0.2, 0.07 and 0.01 are translocated to red bone marrow, liver, kidneys and spleen respectively. The remaining fraction of indium leaving the transfer compartment is assumed to be uniformly distributed throughout all other organs and tissues of the body. Since indium appears to be tenaciously retained in the body,[6] it is appropriate, for the purposes of radiological protection, to assume that indium is retained indefinitely in all organs and tissues.

## 3. Classification of Isotopes for Bone Dosimetry

Indium in the skeleton is primarily associated with bone marrow rather than mineral bone. For the purposes of radiological protection, it is sufficient to assume that the fractions of the systemic burden of indium which are associated with mineral bone and red bone marrow are uniformly distributed throughout those tissues.

# References

1. *ICRP Publication 23, Report of the ICRP Task Group on Reference Man*. Pergamon Press, Oxford, 1975.
2. Smith, G. A., Thomas, R. G. and Scott, J. K. (1960). The Metabolism of Indium after Administration of a Single Dose to the Rat by Intratracheal, Subcutaneous, Intramuscular and Oral Injection. *Health Phys.* **4**, 101–108.
3. Castronovo, F. P. and Wagner, H. N. (1971). Factors Affecting the Toxicity of the Element Indium. *Br. J. Exp. Path.* **52**, 543–559.
4. ICRP Task Group on Lung Dynamics (1966). Deposition and retention models for internal dosimetry of the human respiratory tract. *Health Phys.* **12**, 173–207.
5. Leach, L. J., Scott, J. K., Armstrong, R. D., Steadman, L. T. and Maynard, E. A. *The Inhalation Toxicity of Indium Sesquioxide in the Rat*. AEC Research and Development Report UR-590 (1961).
6. Castronovo, F. P. and Wagner, H. N. (1973). Comparative Toxicity and Pharmacodynamics of Ionic Indium Chloride and Hydrated Indium Oxide. *J. Nucl. Med*, **14**, 677–682.
7. Rayudu, G. V. S., Shirazi, S. P. H. and Fordham, E. W. (1973). Comparison of the Use of $^{52}$Fe and $^{111}$In for Hemopoietic Marrow Scanning. *Int. J. Appl. Radiat. Isot.* **24**, 451–454.

Annual limits on intake, ALI(Bq) and derived air concentrations, DAC(Bq/m$^3$)
(40 h/wk) for isotopes of indium

| Radionuclide | | Oral | Inhalation | |
|---|---|---|---|---|
| | | | Class D | Class W |
| | | $f_1 = 2 \times 10^{-2}$ | $f_1 = 2 \times 10^{-2}$ | $f_1 = 2 \times 10^{-2}$ |
| $^{109}$In | ALI | $7 \times 10^8$ | $2 \times 10^9$ | $2 \times 10^9$ |
| | DAC | — | $7 \times 10^5$ | $1 \times 10^6$ |
| $^{110}$In | ALI | $6 \times 10^8$ | $2 \times 10^9$ | $2 \times 10^9$ |
| (69.1 m) | DAC | — | $7 \times 10^5$ | $9 \times 10^5$ |
| $^{110}$In | ALI | $2 \times 10^8$ | $6 \times 10^8$ | $7 \times 10^8$ |
| (4.9 h) | DAC | — | $3 \times 10^5$ | $3 \times 10^5$ |
| $^{111}$In | ALI | $2 \times 10^8$ | $2 \times 10^8$ | $2 \times 10^8$ |
| | DAC | — | $1 \times 10^5$ | $1 \times 10^5$ |
| $^{112}$In | ALI | $6 \times 10^9$ $(9 \times 10^9)$ ST Wall | $2 \times 10^{10}$ | $3 \times 10^{10}$ |
| | DAC | — | $1 \times 10^7$ | $1 \times 10^7$ |
| $^{113m}$In | ALI | $2 \times 10^9$ | $5 \times 10^9$ | $7 \times 10^9$ |
| | DAC | — | $2 \times 10^6$ | $3 \times 10^6$ |
| $^{114m}$In | ALI | $1 \times 10^7$ $(1 \times 10^7)$ LLI Wall | $2 \times 10^6$ | $4 \times 10^6$ |
| | DAC | — | $1 \times 10^3$ | $2 \times 10^3$ |
| $^{115m}$In | ALI | $5 \times 10^8$ | $2 \times 10^9$ | $2 \times 10^9$ |
| | DAC | — | $7 \times 10^5$ | $7 \times 10^5$ |
| $^{115}$In | ALI | $1 \times 10^6$ | $5 \times 10^4$ | $2 \times 10^5$ |
| | DAC | — | $2 \times 10^1$ | $8 \times 10^1$ |
| $^{116m}$In | ALI | $9 \times 10^8$ | $3 \times 10^9$ | $4 \times 10^9$ |
| | DAC | — | $1 \times 10^6$ | $2 \times 10^6$ |
| $^{117m}$In | ALI | $4 \times 10^8$ | $1 \times 10^9$ | $2 \times 10^9$ |
| | DAC | — | $5 \times 10^5$ | $7 \times 10^5$ |
| $^{117}$In | ALI | $2 \times 10^9$ | $6 \times 10^9$ | $8 \times 10^9$ |
| | DAC | — | $3 \times 10^6$ | $3 \times 10^6$ |
| $^{119m}$In | ALI | $1 \times 10^9$ $(2 \times 10^9)$ ST Wall | $5 \times 10^9$ | $5 \times 10^9$ |
| | DAC | — | $2 \times 10^6$ | $2 \times 10^6$ |

# METABOLIC DATA FOR XENON

No metabolic model is proposed for xenon. As explained in Chapter 8 of this report, exposure in a cloud of a radioactive noble gas is limited by external irradiation, since dose-equivalent rates from gas absorbed in tissue or contained in the lungs will be negligible in comparison with the dose-equivalent rates to tissues from external irradiation. The recommended DACs for xenon are, therefore, based on consideration of external irradiation only.

Derived air concentrations, DAC(Bq/m$^3$) (40 h/wk) for isotopes of xenon

| Radionuclide | Semi-infinite Cloud | 1 000 m$^3$ room | 500 m$^3$ room | 100 m$^3$ room |
|---|---|---|---|---|
| $^{120}$Xe | $4 \times 10^5$ | $7 \times 10^6$ | $9 \times 10^6$ | $2 \times 10^7$ |
| $^{121}$Xe | $8 \times 10^4$ | $2 \times 10^6$<br>$(2 \times 10^6)$<br>Skin | $2 \times 10^6$<br>$(2 \times 10^6)$<br>Skin | $2 \times 10^6$<br>$(4 \times 10^6)$<br>Skin |
| $^{122}$Xe | $3 \times 10^6$ | $4 \times 10^7$ | $5 \times 10^7$ | $9 \times 10^7$ |
| $^{123}$Xe | $2 \times 10^5$ | $5 \times 10^6$ | $6 \times 10^6$<br>$(7 \times 10^6)$<br>Skin | $6 \times 10^6$<br>$(1 \times 10^7)$<br>Skin |
| $^{125}$Xe | $6 \times 10^5$ | $1 \times 10^7$ | $1 \times 10^7$ | $2 \times 10^7$ |
| $^{127}$Xe | $5 \times 10^5$ | $1 \times 10^7$ | $1 \times 10^7$ | $2 \times 10^7$ |
| $^{129m}$Xe | $7 \times 10^6$ | $1 \times 10^7$<br>$(5 \times 10^7)$<br>Skin | $1 \times 10^7$<br>$(7 \times 10^7)$<br>Skin | $1 \times 10^7$<br>$(1 \times 10^8)$<br>Skin |
| $^{131m}$Xe | $1 \times 10^7$<br>$(2 \times 10^7)$<br>Skin | $2 \times 10^7$<br>$(1 \times 10^8)$<br>Skin | $2 \times 10^7$<br>$(2 \times 10^8)$<br>Skin | $2 \times 10^7$<br>$(3 \times 10^8)$<br>Skin |
| $^{133m}$Xe | $5 \times 10^6$ | $8 \times 10^6$<br>$(7 \times 10^7)$<br>Skin | $8 \times 10^6$<br>$(8 \times 10^7)$<br>Skin | $8 \times 10^6$<br>$(1 \times 10^8)$<br>Skin |
| $^{133}$Xe | $4 \times 10^6$ | $2 \times 10^7$<br>$(8 \times 10^7)$<br>Skin | $2 \times 10^7$<br>$(1 \times 10^8)$<br>Skin | $2 \times 10^7$<br>$(2 \times 10^8)$<br>Skin |
| $^{135m}$Xe | $3 \times 10^5$ | $7 \times 10^6$ | $9 \times 10^6$ | $1 \times 10^7$<br>$(2 \times 10^7)$<br>Skin |
| $^{135}$Xe | $5 \times 10^5$ | $4 \times 10^6$<br>$(1 \times 10^7)$<br>Skin | $4 \times 10^6$<br>$(2 \times 10^7)$<br>Skin | $4 \times 10^6$<br>$(3 \times 10^7)$<br>Skin |
| $^{138}$Xe | $1 \times 10^5$ | $2 \times 10^6$<br>$(3 \times 10^6)$<br>Skin | $2 \times 10^6$<br>$(4 \times 10^6)$<br>Skin | $2 \times 10^6$<br>$(7 \times 10^6)$<br>Skin |

# METABOLIC DATA FOR BARIUM

## 1. Metabolism

Data from Reference Man[1]

| | |
|---|---|
| Barium content of the body | 22 mg |
| of bone | 20 mg |
| Daily intake in food and fluids | 0.75 mg |

## 2. Metabolic Model

### (a) Uptake to blood

The fractional absorption of barium from most diets[1,2] and from simulated fallout[3] is less than is the fractional absorption of strontium.[1] In this report $f_1$ is assumed to be 0.1 for all compounds of barium.

### (b) Inhalation classes

The ICRP Task Group on Lung Dynamics[4] assigned all compounds of barium to inhalation class W. However, experiments on dogs[5,6] with $^{131}BaSO_4$ and $^{140}BaCl_2$ have shown that barium is cleared more rapidly from the lung than this classification would suggest. In this report all compounds of barium are assigned to inhalation class D.

| Inhalation Class | $f_1$ |
|---|---|
| D | 0.1 |
| W | — |
| Y | — |

### (c) Distribution and retention

A very comprehensive model for the retention of barium in adults has been developed by the Task Group on Alkaline Earth Metabolism in Adult Man.[7] The total number of spontaneous nuclear transformations in soft tissue, cortical bone and trabecular bone during the 50 years following the introduction of 1Bq of a radioactive isotope of barium into the transfer compartment can be derived from the retention functions given in the Task Group report.[7]

### (d) Behaviour of daughters

$^{127}Ba$ decays via $^{127}Cs$ to $^{127}Xe$ which has a radioactive half-life of 36 days and is assumed to escape from the body without decaying.

## 3. Classification of Isotopes for Bone Dosimetry

As discussed in Chapter 7, isotopes of the alkaline earths with radioactive half-lives of greater than 15 days are assumed to be uniformly distributed throughout the volume of mineral bone, whereas isotopes with radioactive half-lives of less than 15 days are assumed to be distributed in a thin layer over bone surfaces. Thus, when deriving values of absorbed fractions (see Table 7.4, Chapter 7) $^{133}Ba$ is assumed to be distributed throughout the volume of mineral bone and all other radioactive isotopes of barium are assumed to be distributed over bone surfaces at all times following their deposition in the skeleton.

## References

1. *ICRP Publication 23, Report of the ICRP Task Group on Reference Man.* Pergamon Press, Oxford, 1975.
2. Underwood, E. C. *Trace Elements in Human and Animal Nutrition.* Academic Press, London, 1971, pp. 431–432.
3. LeRoy, G. V., Rust, J. H. and Hasterlik, R. J. (1966). The Consequences of Ingestion by Man of Real and Simulated Fall-Out. *Health Phys.* **12**, 449–473.
4. ICRP Task Group on Lung Dynamics (1966). Deposition and retention models for internal dosimetry of the human respiratory tract. *Health Phys.* **12**, 173–207.
5. Morrow, P. E., Gibb, F. R. and Johnson, L. (1964). Clearance of Insoluble Dust from the Lower Respiratory Tract. *Health Phys.* **10**, 543–555.
6. Cuddihy, R. G. and Griffith, W. C. (1972). A Biological Model Describing Tissue Distribution and Whole-Body Retention of Barium and Lanthanum in Beagle Dogs after Inhalation and Lavage. *Health Phys.* **23**, 621–633.
7. *ICRP Publication 20, Report of the ICRP Task Group on Alkaline Earth Metabolism in Adult Man.* Pergamon Press, Oxford, 1972.

Annual limits on intake, ALI(Bq) and derived air concentrations,
DAC(Bq/m$^3$) (40 h/wk) for isotopes of barium

| Radionuclide | | Oral | Inhalation Class D |
|---|---|---|---|
| | | $f_1 = 1 \times 10^{-1}$ | $f_1 = 1 \times 10^{-1}$ |
| $^{126}$Ba | ALI | $2 \times 10^8$ | $6 \times 10^8$ |
| | DAC | — | $2 \times 10^5$ |
| $^{128}$Ba | ALI | $2 \times 10^7$ | $7 \times 10^7$ |
| | DAC | — | $3 \times 10^4$ |
| $^{131m}$Ba | ALI | $1 \times 10^{10}$ $(2 \times 10^{10})$ ST Wall | $5 \times 10^{10}$ |
| | DAC | — | $2 \times 10^7$ |
| $^{131}$Ba | ALI | $1 \times 10^8$ | $3 \times 10^8$ |
| | DAC | — | $1 \times 10^5$ |
| $^{133m}$Ba | ALI | $9 \times 10^7$ $(1 \times 10^8)$ LLI Wall | $3 \times 10^8$ |
| | DAC | — | $1 \times 10^5$ |
| $^{133}$Ba | ALI | $6 \times 10^7$ | $3 \times 10^7$ |
| | DAC | — | $1 \times 10^4$ |
| $^{135m}$Ba | ALI | $1 \times 10^8$ | $4 \times 10^8$ |
| | DAC | — | $2 \times 10^5$ |
| $^{139}$Ba | ALI | $5 \times 10^8$ | $1 \times 10^9$ |
| | DAC | — | $5 \times 10^5$ |
| $^{140}$Ba | ALI | $2 \times 10^7$ $(2 \times 10^7)$ LLI Wall | $5 \times 10^7$ |
| | DAC | — | $2 \times 10^4$ |
| $^{141}$Ba | ALI | $9 \times 10^8$ | $3 \times 10^9$ |
| | DAC | — | $1 \times 10^6$ |
| $^{142}$Ba | ALI | $2 \times 10^9$ | $5 \times 10^9$ |
| | DAC | — | $2 \times 10^6$ |

# METABOLIC DATA FOR RHENIUM

## 1. Metabolism

No data are given in Reference Man[1] for rhenium.

## 2. Metabolic Model

### (a) Uptake to blood

There appears to be no information concerning the uptake of rhenium from the gastrointestinal tract of any mammalian species. The scanty biological information which is available[2,3] suggests that the metabolic behaviour of the element is rather similar to that of technetium. In this report $f_1$ is assumed to be 0.8 for all compounds of rhenium.

### (b) Inhalation classes

The ICRP Task Group on Lung Dynamics[4] assigned oxides, hydroxides, halides and nitrates of rhenium to inhalation class W and all other compounds of the element to inhalation class D. In the absence of any relevant experimental data this classification is adopted here.

| Inhalation Class | $f_1$ |
|:---:|:---:|
| D | 0.8 |
| W | 0.8 |
| Y | — |

### (c) Distribution and retention

The limited information available[2,3] suggests that the metabolic behaviour of rhenium is similar to that of technetium. In this report the metabolic model used for technetium is also used for rhenium.

It is assumed that of rhenium leaving the transfer compartment 0.04 is translocated to the thyroid where it is retained with a biological half-life of 0.5 days. Further fractions 0.1 and 0.03 are assumed to be translocated to the stomach wall and liver respectively. The remaining fraction of rhenium leaving the transfer compartment is assumed to be uniformly distributed throughout all organs and tissues of the body other than the thyroid, stomach wall and liver. Of rhenium translocated to any organ or tissue of the body other than the thyroid, fractions 0.75, 0.20 and 0.05 are assumed to be retained with biological half-lives of 1.6, 3.7 and 22 days respectively. For rhenium the biological half-life in the transfer compartment is taken to be 0.02 day.

## 3. Classification of Isotopes for Bone Dosimetry

Rhenium is assumed to be uniformly distributed throughout all organs and tissues of the body other than the thyroid, stomach wall and liver. Therefore, a classification of isotopes of the element for the purposes of bone dosimetry is not required.

# References

1. *ICRP Publication 23, Report of the ICRP Task Group on Reference Man.* Pergamon Press, Oxford, 1975.
2. Durbin, P. W., Scott, K. G. and Hamilton, J. G. (1957). The Distribution of Radioisotopes of some Heavy Metals in the Rat. *University of California Publications in Pharmacology* **3**, 1-34.
3. Durbin, P. W. (1960). Metabolic Characteristics within a Chemical Family. *Health Phys.* **2**, 225-238.
4. ICRP Task Group on Lung Dynamics (1966). Deposition and retention models for internal dosimetry of the human respiratory tract. *Health Phys.* **12**, 173-207.

Annual limits on intake, ALI(Bq) and derived air concentrations, DAC(Bq/m$^3$) (40 h/wk) for isotopes of rhenium

| Radionuclide | | Oral | Inhalation | |
| --- | --- | --- | --- | --- |
| | | | Class D | Class W |
| | | $f_1 = 8 \times 10^{-1}$ | $f_1 = 8 \times 10^{-1}$ | $f_1 = 8 \times 10^{-1}$ |
| $^{177}$Re | ALI | $4 \times 10^9$ $(4 \times 10^9)$ ST Wall | $1 \times 10^{10}$ | $1 \times 10^{10}$ |
| | DAC | — | $4 \times 10^6$ | $5 \times 10^6$ |
| $^{178}$Re | ALI | $3 \times 10^9$ $(4 \times 10^9)$ ST Wall | $1 \times 10^{10}$ | $1 \times 10^{10}$ |
| | DAC | — | $4 \times 10^6$ | $4 \times 10^6$ |
| $^{181}$Re | ALI | $2 \times 10^8$ | $3 \times 10^8$ | $3 \times 10^8$ |
| | DAC | — | $1 \times 10^5$ | $1 \times 10^5$ |
| $^{182}$Re (12.7h) | ALI | $3 \times 10^8$ | $5 \times 10^8$ | $6 \times 10^8$ |
| | DAC | — | $2 \times 10^5$ | $2 \times 10^5$ |
| $^{182}$Re (64.0h) | ALI | $5 \times 10^7$ | $9 \times 10^7$ | $8 \times 10^7$ |
| | DAC | — | $4 \times 10^4$ | $3 \times 10^4$ |
| $^{184m}$Re | ALI | $8 \times 10^7$ | $1 \times 10^8$ | $2 \times 10^7$ |
| | DAC | — | $5 \times 10^4$ | $7 \times 10^3$ |
| $^{184}$Re | ALI | $9 \times 10^7$ | $1 \times 10^8$ | $5 \times 10^7$ |
| | DAC | — | $5 \times 10^4$ | $2 \times 10^4$ |
| $^{186m}$Re | ALI | $5 \times 10^7$ $(6 \times 10^7)$ ST Wall | $6 \times 10^7$ $(8 \times 10^7)$ ST Wall | $6 \times 10^6$ |
| | DAC | — | $3 \times 10^4$ | $2 \times 10^3$ |
| $^{186}$Re | ALI | $7 \times 10^7$ | $1 \times 10^8$ | $6 \times 10^7$ |
| | DAC | — | $4 \times 10^4$ | $3 \times 10^4$ |
| $^{187}$Re | ALI | $2 \times 10^{10}$ | $3 \times 10^{10}$ $(3 \times 10^{10})$ ST Wall | $4 \times 10^9$ |
| | DAC | — | $1 \times 10^7$ | $2 \times 10^6$ |
| $^{188m}$Re | ALI | $3 \times 10^9$ | $5 \times 10^9$ | $5 \times 10^9$ |
| | DAC | — | $2 \times 10^6$ | $2 \times 10^6$ |
| $^{188}$Re | ALI | $6 \times 10^7$ | $1 \times 10^8$ | $1 \times 10^8$ |
| | DAC | — | $4 \times 10^4$ | $4 \times 10^4$ |
| $^{189}$Re | ALI | $1 \times 10^8$ | $2 \times 10^8$ | $2 \times 10^8$ |
| | DAC | — | $8 \times 10^4$ | $7 \times 10^4$ |

# METABOLIC DATA FOR OSMIUM

## 1. Metabolism

No data are given in Reference Man[1] for Osmium.

## 2. Metabolic Model

(a) *Uptake to blood*

There appears to be no information available concerning the uptake of osmium from the gastrointestinal tract. Chemically osmium resembles iridium and $f_1$ has, therefore, been taken to be 0.01 for all compounds of the element.

(b) *Inhalation classes*

The ICRP Task Group on Lung Dynamics[2] assigned oxides and hydroxides of osmium to inhalation class Y, halides and nitrates to inhalation class W and all other compounds of the element to inhalation class D. In the absence of any relevant experimental data and by analogy with iridium the recommendations of the Task Group[2] are adopted in this report.

| Inhalation Class | $f_1$ |
|---|---|
| D | 0.01 |
| W | 0.01 |
| Y | 0.01 |

(c) *Distribution and retention*

The available data on animals[3,4] indicate that the metabolic behaviour of osmium does not differ markedly from that of the other platinum metals. The metabolic model used for iridium is, therefore, also used for osmium.

In this report it is assumed that of osmium leaving the transfer compartment fractions 0.2, 0.04 and 0.02 are translocated to liver, kidney and spleen respectively. A further fraction, 0.54, is assumed to be uniformly distributed throughout all other organs and tissues of the body. The remaining fraction of osmium leaving the transfer compartment is assumed to go directly to excreta. Of osmium deposited in any organ or tissue of the body fractions 0.2 and 0.8 are assumed to be retained with biological half-lives of 8 and 200 days respectively.

## 3. Classification of Isotopes for Bone Dosimetry

Osmium is assumed to be uniformly distributed throughout all organs and tissues of the body other than the liver, kidney and spleen. Therefore, a classification of isotopes of the element for the purposes of bone dosimetry is not required.

# References

1. *ICRP Publication 23, Report of the ICRP Task Group on Reference Man.* Pergamon Press, Oxford, 1975.
2. ICRP Task Group Report on Lung Dynamics (1966). Deposition and retention models for internal dosimetry of the human respiratory tract. *Health Phys.* **12**, 173–207.
3. Durbin, P. W., Scott, K. G. and Hamilton, J. G. (1957). The Distribution of Radioisotopes of Some Heavy Metals in the Rat. *University of California Publication in Pharmacology* **3**, 1–34.
4. Durbin, P. W. (1960). Metabolic Characteristics within a Chemical Family. *Health Phys.* **2**, 225–238.

Annual limits on intake, ALI(Bq) and derived air concentrations, DAC(Bq/m$^3$)
(40 h/wk) for isotopes of osmium

| Radionuclide | | Oral | Inhalation | | |
| --- | --- | --- | --- | --- | --- |
| | | | Class D | Class W | Class Y |
| | | $f_1 = 1 \times 10^{-2}$ | $f_1 = 1 \times 10^{-2}$ | $f_1 = 1 \times 10^{-2}$ | $f_1 = 1 \times 10^{-2}$ |
| $^{180}$Os | ALI | $4 \times 10^9$ | $1 \times 10^{10}$ | $2 \times 10^{10}$ | $2 \times 10^{10}$ |
| | DAC | — | $6 \times 10^6$ | $7 \times 10^6$ | $7 \times 10^6$ |
| $^{181}$Os | ALI | $5 \times 10^8$ | $2 \times 10^9$ | $2 \times 10^9$ | $2 \times 10^9$ |
| | DAC | — | $7 \times 10^5$ | $7 \times 10^5$ | $7 \times 10^5$ |
| $^{182}$Os | ALI | $8 \times 10^7$ | $2 \times 10^8$ | $2 \times 10^8$ | $1 \times 10^8$ |
| | DAC | — | $9 \times 10^4$ | $7 \times 10^4$ | $6 \times 10^4$ |
| $^{185}$Os | ALI | $9 \times 10^7$ | $2 \times 10^7$ | $3 \times 10^7$ | $3 \times 10^7$ |
| | DAC | — | $8 \times 10^3$ | $1 \times 10^4$ | $1 \times 10^4$ |
| $^{189m}$Os | ALI | $3 \times 10^9$ | $9 \times 10^9$ | $8 \times 10^9$ | $6 \times 10^9$ |
| | DAC | — | $4 \times 10^6$ | $3 \times 10^6$ | $3 \times 10^6$ |
| $^{191m}$Os | ALI | $5 \times 10^8$ | $1 \times 10^9$ | $8 \times 10^8$ | $7 \times 10^8$ |
| | DAC | — | $4 \times 10^5$ | $3 \times 10^5$ | $3 \times 10^5$ |
| $^{191}$Os | ALI | $8 \times 10^7$ $(9 \times 10^7)$ LLI Wall | $8 \times 10^7$ | $6 \times 10^7$ | $5 \times 10^7$ |
| | DAC | — | $3 \times 10^4$ | $2 \times 10^4$ | $2 \times 10^4$ |
| $^{193}$Os | ALI | $6 \times 10^7$ $(6 \times 10^7)$ LLI Wall | $2 \times 10^8$ | $1 \times 10^8$ | $1 \times 10^8$ |
| | DAC | — | $7 \times 10^4$ | $5 \times 10^4$ | $4 \times 10^4$ |
| $^{194}$Os | ALI | $2 \times 10^7$ $(2 \times 10^7)$ LLI Wall | $2 \times 10^6$ | $2 \times 10^6$ | $3 \times 10^5$ |
| | DAC | — | $6 \times 10^2$ | $9 \times 10^2$ | $1 \times 10^2$ |

# METABOLIC DATA FOR IRIDIUM

## 1. Metabolism

No data are given in Reference Man[1] for Iridium.

## 2. Metabolic model

(a) *Uptake to blood*

The uptake of $Na_2 {}^{192}IrCl_6$ from the gastrointestinal tract has been measured in several mammalian species.[2] The results indicate a fractional absorption of about 0.01 for iridium administered in this form. The absorption of carrier-free iridium has been reported as about 0.1[3] but a later report[4] indicated a value of less than $10^{-3}$. In this report $f_1$ is taken to be 0.01 for all compounds of iridium.

(b) *Inhalation classes*

The ICRP Task Group on Lung Dynamics[5] assigned oxides and hydroxides of iridium to inhalation class Y, halides and nitrates to inhalation class W and all other compounds of the element to inhalation class D. In man an accidentally inhaled compound of iridium was cleared rapidly from the lungs[6] and in the rat metallic iridium was found to behave as a class W material.[7]

In this report the recommendations of the Task Group[5] are adopted for compounds of irridium. Metallic iridium is assigned to inhalation class W.

| Inhalation Class | $f_1$ |
|---|---|
| D | 0.01 |
| W | 0.01 |
| Y | 0.01 |

(c) *Distribution and retention*

Experiments on rats[2,8] indicate that the concentration of iridium in the liver, kidney and spleen at any time following either intravenous or intraperitoneal injection is about an order of magnitude higher than the average whole body concentration. It has further been shown[2] that the retention of iridium in each of these organs parallels the whole body retention.

The whole body retention of iridium is very similar in mice, rats, monkeys and dogs.[2] Data from experiments on these four species suggest that a retention function appropriate for man is

$$R(t) = 0.2e^{-0.693t/0.3} + 0.15e^{-0.693t/8} + 0.65e^{-0.693t/200}$$

In this report it is assumed that of iridium leaving the transfer compartment fractions 0.2, 0.04 and 0.02 are translocated to liver, kidney and spleen respectively. A further fraction, 0.54, is assumed to be uniformly distributed throughout all other organs and tissues of the body. The remaining fraction of iridium leaving the transfer compartment is assumed to go directly to excreta. Of iridium deposited in any organ or tissue of the body fractions 0.2 and 0.8 are assumed to be retained with biological half-lives of 8 and 200 days respectively.

## 3. Classification of Isotopes for Bone Dosimetry

Iridium is assumed to be uniformly distributed throughout all organs and tissues of the body other than the liver, kidney and spleen. Therefore, a classification of isotopes of the element for the purposes of bone dosimetry is not required.

# References

1. *ICRP Publication 23, Report of the ICRP Task Group on Reference Man.* Pergamon Press, Oxford, 1975.
2. Furchner, J. E., Richmond, C. R. and Drake, G. A. (1971). Comparative Metabolism of Radionuclides in Mammals - V. Retention of $^{192}$Ir in the Mouse, Rat, Monkey and Dog. *Health Phys.* **20**, 375–382.
3. Hamilton, J. G. (1951). The Metabolic Properties of Various Materials. *University of California Radiation Laboratory Report.* URCRL-1437.
4. Durbin, P. W., Scott, K. G. and Hamilton, J. G. (1957). The Distribution of Radioisotopes of Some Heavy Metals in the Rat. *University of California Publications in Pharmacology* **3**, 1–34.
5. ICRP Task Group on Lung Dynamics (1966). Deposition and retention models for internal dosimetry of the human respiratory tract. *Health Phys.* **12**, 173–207.
6. Brodsky, A., Schubert, J., Yaniv, S., Lamson, K., Wald, N., Wechsler, R., Gumerman, L. and Caldwell, R. (1967). Deposition and Retention of $^{192}$Ir in the Lung after an Inhalation Incident. *Health Phys.* **13**, p. 938.
7. Casarett, L. J., Bless, S., Katz, R. and Scott, J. K. (1960). Retention and Fate of Iridium-192 in rats following Inhalation. *J. Am. Ind. Hyg. Assoc.* **21**, 414–418.
8. Durbin, P. W. (1960). Metabolic Characteristics within a Chemical Family. *Health Phys.* **2**, 225–238.

Annual limits on intake, ALI(Bq) and derived air concentrations, DAC(Bq/m$^3$)
(40 h/wk) for isotopes of iridium

| Radionuclide | | Oral | Inhalation | | |
|---|---|---|---|---|---|
| | | | Class D | Class W | Class Y |
| | | $f_1 = 1 \times 10^{-2}$ | $f_1 = 1 \times 10^{-2}$ | $f_1 = 1 \times 10^{-2}$ | $f_1 = 1 \times 10^{-2}$ |
| $^{182}$Ir | ALI | $2 \times 10^9$ $(2 \times 10^9)$ ST Wall | $5 \times 10^9$ | $6 \times 10^9$ | $5 \times 10^9$ |
| | DAC | — | $2 \times 10^6$ | $2 \times 10^6$ | $2 \times 10^6$ |
| $^{184}$Ir | ALI | $3 \times 10^8$ | $9 \times 10^8$ | $1 \times 10^9$ | $1 \times 10^9$ |
| | DAC | — | $4 \times 10^5$ | $5 \times 10^5$ | $4 \times 10^5$ |
| $^{185}$Ir | ALI | $2 \times 10^8$ | $5 \times 10^8$ | $4 \times 10^8$ | $4 \times 10^8$ |
| | DAC | — | $2 \times 10^5$ | $2 \times 10^5$ | $2 \times 10^5$ |
| $^{186}$Ir | ALI | $9 \times 10^7$ | $3 \times 10^8$ | $2 \times 10^8$ | $2 \times 10^8$ |
| | DAC | — | $1 \times 10^5$ | $1 \times 10^5$ | $9 \times 10^4$ |
| $^{187}$Ir | ALI | $4 \times 10^8$ | $1 \times 10^9$ | $1 \times 10^9$ | $1 \times 10^9$ |
| | DAC | — | $5 \times 10^5$ | $5 \times 10^5$ | $4 \times 10^5$ |
| $^{188}$Ir | ALI | $7 \times 10^7$ | $2 \times 10^8$ | $1 \times 10^8$ | $1 \times 10^8$ |
| | DAC | — | $7 \times 10^4$ | $5 \times 10^4$ | $5 \times 10^4$ |
| $^{189}$Ir | ALI | $2 \times 10^8$ $(2 \times 10^8)$ LLI Wall | $2 \times 10^8$ | $1 \times 10^8$ | $1 \times 10^8$ |
| | DAC | — | $7 \times 10^4$ | $6 \times 10^4$ | $6 \times 10^4$ |
| $^{190m}$Ir | ALI | $6 \times 10^9$ | $7 \times 10^9$ | $8 \times 10^9$ | $7 \times 10^9$ |
| | DAC | — | $3 \times 10^6$ | $3 \times 10^6$ | $3 \times 10^6$ |
| $^{190}$Ir | ALI | $4 \times 10^7$ | $3 \times 10^7$ | $4 \times 10^7$ | $3 \times 10^7$ |
| | DAC | — | $1 \times 10^4$ | $2 \times 10^4$ | $1 \times 10^4$ |
| $^{192m}$Ir | ALI | $1 \times 10^8$ | $3 \times 10^6$ | $8 \times 10^6$ | $6 \times 10^5$ |
| | DAC | — | $1 \times 10^3$ | $3 \times 10^3$ | $2 \times 10^2$ |
| $^{192}$Ir | ALI | $4 \times 10^7$ | $1 \times 10^7$ | $1 \times 10^7$ | $8 \times 10^6$ |
| | DAC | — | $4 \times 10^3$ | $6 \times 10^3$ | $3 \times 10^3$ |
| $^{194m}$Ir | ALI | $2 \times 10^7$ | $3 \times 10^6$ | $6 \times 10^6$ | $4 \times 10^6$ |
| | DAC | — | $1 \times 10^3$ | $3 \times 10^3$ | $2 \times 10^3$ |
| $^{194}$Ir | ALI | $4 \times 10^7$ | $1 \times 10^8$ | $8 \times 10^7$ | $7 \times 10^7$ |
| | DAC | — | $5 \times 10^4$ | $3 \times 10^4$ | $3 \times 10^4$ |
| $^{195m}$Ir | ALI | $3 \times 10^8$ | $9 \times 10^8$ | $1 \times 10^9$ | $8 \times 10^8$ |
| | DAC | — | $4 \times 10^5$ | $4 \times 10^5$ | $3 \times 10^5$ |
| $^{195}$Ir | ALI | $6 \times 10^8$ | $2 \times 10^9$ | $2 \times 10^9$ | $2 \times 10^9$ |
| | DAC | — | $6 \times 10^5$ | $8 \times 10^5$ | $7 \times 10^5$ |

# METABOLIC DATA FOR GOLD

## 1. Metabolism

Data from Reference Man[1]
Gold content of the body                    $<9.8$ mg
No data concerning the normal daily
intake of gold in food and and fluids are
given in Reference Man.

## 2. Metabolic Model

(a) *Uptake to blood*

The fractional absorption of various salts of gold from the gastrointestinal tract has been found to vary from 0.03 to 0.13.[2-4] In this report $f_1$ is taken to be 0.1 for all compounds of gold.

(b) *Inhalation classes*

The ICRP Task Group on Lung Dynamics[5] assigned oxides and hydroxides of gold to inhalation class Y, halides and nitrates to inhalation class W and all other compounds of the element to inhalation class D. In the absence of any relevant experimental data this classification is adopted here.

| Inhalation Class | $f_1$ |
|---|---|
| D | 0.1 |
| W | 0.1 |
| Y | 0.1 |

(c) *Distribution and retention*

Studies on patients with active classic rheumatoid arthritis[6] indicate that gold injected as gold sodium thiomalate is retained in the body with a biological half-life of about 3 days. Similar biological half-lives have been found in other studies with gold salts.[7]

It should be noted that, because of the rapid urinary excretion of gold salts, average absorbed dose rates to the walls of the urinary bladder will be considerably greater than average absorbed dose rates to any other organ or tissue of the body.

In this report it is assumed that gold entering the transfer compartment is instantaneously uniformly distributed throughout all organs and tissues of the body, where it is retained with a biological half-life of 3 days. It is also assumed that the concentration of a radioactive isotope of gold in urine contained in the bladder is 10 times the concentration in any other tissue at all times after exposure to that isotope and that the urine content of the bladder is 200 cm$^3$.

## 3. Classification of Isotopes for Bone Dosimetry

Because gold is assumed to be uniformly distributed in body tissues, a classification of isotopes of the element for the purpose of bone dosimetry is not required.

## References

1. *ICRP Publication 23, Report of the ICRP Task Group on Reference Man.* Pergamon Press, Oxford, 1975.
2. Sylvia, A. J., Fleshman, D. G. and Shore, B. (1973). The Effects of Penicillamine on the Body Burdens of Several Heavy Metals. *Health Phys.* **24**, 535–539.
3. Chertok, R. J. and Lake, S. (1971). Biological Availability of Radionuclides Produced by the Plowshare Event Schooner-II Retention and Excretion Rates in Peccaries after a Single Oral Dose of Debris. *Health Phys.* **20**, 325–330.
4. Kleinsorge, H. (1967). Die Resorption Therapeutisch Anwendbarer Goldsalze und Goldsole. *Arzneim.-Forsch.* **17**, 100–102.
5. ICRP Task Group on Lung Dynamics (1966). Deposition and retention models for internal dosimetry of the human respiratory tract. *Health Phys.* **12**, 173–207.
6. Mascarenhas, B. R., Oranda, J. L. and Freyberg, R. H. (1972). Gold Metabolism in Patients with Rheumatoid Arthritis Treated with Gold Compounds—Reinvestigated. *Arthritis Rheum.* **15**, 391–402.
7. *ICRP Publication 10, Report of Committee IV on Evaluation of Radiation Doses to Body Tissues from Internal Contamination due to Occupational Exposure.* Pergamon Press, Oxford, 1968.

Annual limits on intake, ALI(Bq) and derived air concentrations, DAC(Bq/m$^3$)
(40 h/wk) for isotopes of gold

| Radionuclide | | Oral | Inhalation | | |
| --- | --- | --- | --- | --- | --- |
| | | | Class D | Class W | Class Y |
| | | $f_1 = 1 \times 10^{-1}$ | $f_1 = 1 \times 10^{-1}$ | $f_1 = 1 \times 10^{-1}$ | $f_1 = 1 \times 10^{-1}$ |
| $^{193}$Au | ALI | $3 \times 10^8$ | $5 \times 10^8$ | $6 \times 10^8$ | $6 \times 10^8$ |
| | DAC | — | $2 \times 10^5$ | $3 \times 10^5$ | $3 \times 10^5$ |
| $^{194}$Au | ALI | $9 \times 10^7$ | $1 \times 10^8$ | $2 \times 10^8$ | $2 \times 10^8$ |
| | DAC | — | $6 \times 10^4$ | $7 \times 10^4$ | $7 \times 10^4$ |
| $^{195}$Au | ALI | $2 \times 10^8$ | $1 \times 10^8$ ($2 \times 10^8$) BLAD Wall | $5 \times 10^7$ | $2 \times 10^7$ |
| | DAC | — | $5 \times 10^4$ | $2 \times 10^4$ | $7 \times 10^3$ |
| $^{198m}$Au | ALI | $3 \times 10^7$ | $3 \times 10^7$ ($4 \times 10^7$) BLAD Wall | $4 \times 10^7$ | $4 \times 10^7$ |
| | DAC | — | $1 \times 10^4$ | $2 \times 10^4$ | $2 \times 10^4$ |
| $^{198}$Au | ALI | $4 \times 10^7$ | $4 \times 10^7$ ($6 \times 10^7$) BLAD Wall | $6 \times 10^7$ | $6 \times 10^7$ |
| | DAC | — | $2 \times 10^4$ | $2 \times 10^4$ | $2 \times 10^4$ |
| $^{199}$Au | ALI | $1 \times 10^8$ ($1 \times 10^8$) LLI Wall | $1 \times 10^8$ ($1 \times 10^8$) BLAD Wall | $1 \times 10^8$ | $1 \times 10^8$ |
| | DAC | — | $4 \times 10^4$ | $5 \times 10^4$ | $5 \times 10^4$ |
| $^{200m}$Au | ALI | $4 \times 10^7$ | $7 \times 10^7$ | $8 \times 10^7$ | $9 \times 10^7$ |
| | DAC | — | $3 \times 10^4$ | $3 \times 10^4$ | $4 \times 10^4$ |
| $^{200}$Au | ALI | $1 \times 10^9$ | $1 \times 10^9$ | $2 \times 10^9$ | $3 \times 10^9$ |
| | DAC | — | $6 \times 10^5$ | $1 \times 10^6$ | $1 \times 10^6$ |
| $^{201}$Au | ALI | $3 \times 10^9$ ($3 \times 10^9$) ST Wall | $5 \times 10^9$ | $8 \times 10^9$ | $8 \times 10^9$ |
| | DAC | — | $2 \times 10^6$ | $3 \times 10^6$ | $3 \times 10^6$ |

# METABOLIC DATA FOR MERCURY

## 1. Metabolism

Data from Reference Man.[1]

| | |
|---|---|
| Mercury content of soft tissues | 13 mg |
| Daily intake in food and fluids | 0.015 mg |

## 2. Metabolic Models

### 2.1 Inorganic Compounds

(a) *Uptake to blood*

The uptake of elemental mercury from the gastrointestinal tract is very limited[2] and experiments on rats[3] suggest that less than $10^{-4}$ of ingested elemental mercury is absorbed.

Absorption of inorganic compounds of mercury from the gastrointestinal tract has been reviewed by Nordberg and Sherfving.[2] Studies on mice suggests that the fractional absorption of mercuric chloride is less than 0.02. However, data from acute cases of poisoning in man suggest a fractional absorption of at least 0.08 for mercury ingested in this form. The difference between the two figures is probably explained by the disruptive effect of mercuric chloride on the permeability barriers of the gastrointestinal tract and use of the lower figure is, therefore, probably more appropriate in radiological protection.

It is anticipated that the fractional absorption of mercurous compounds from the gastrointestinal tract will, in general, be less than the fractional absorption of mercuric compounds.[2]

In this report $f_1$ is taken as 0.02 for all inorganic compounds of mercury.

(b) *Inhalation classes*

Animal studies on the deposition and retention of mercury vapour in the lung have been reviewed by Nordberg and Sherfving.[2] Absorption of mercury vapour from the lung is typically more than 0.5 of that inhaled. In man an average of 0.74 of mercury inhaled is retained in the lungs[4] with a biological half-life of 1.7 days. Most of this mercury is translocated to body tissues, since only about 0.07 of the total body burden is lost as vapour in the expired air.

In this report it is assumed that 0.7 of mercury entering the lung as mercury vapour is deposited there and that following deposition this fraction is translocated to blood with a biological half-life of 1.7 days. The metabolism of this mercury, subsequent to its entry into the blood, is assumed to be identical to that of inorganic compounds of the element.

The ICRP Task Group on Lung Dynamics[5] assigned oxides, hydroxides, halides, nitrates and sulphides of mercury to inhalation class W and sulphates of the element to inhalation class D. This classification is supported by experiments in which dogs inhaled $^{203}HgO$[6] and by studies on accidentally exposed humans[13] and is adopted here.

| Inhalation Class | $f_1$ |
| --- | --- |
| D | 0.02 |
| W | 0.02 |
| Y | — |

(c) *Distribution and retention*

The organ with the highest concentration of inorganic mercury is usually the kidney[1,2] which has, typically about 15 times the average whole body concentration.[1] However, at high levels of exposure to inorganic mercury, concentrations in the thyroid and pituitary glands may exceed the concentration in the kidney.[7]

Experiments involving the oral administration of inorganic and methyl mercury to a considerable number of individuals[7] have shown that, over the first 250 days after ingestion, inorganic mercury has a biological half-life in the whole body of $42 \pm 3$ days while methyl mercury has a biological half-life in the whole body of $76 \pm 3$ days. However, these half-lives are not compatible with the total body content and daily intake of mercury given for Reference Man.[1] Nor are they compatible with the observation that mercury is still detectable in the urine 6 years after the last known exposure to the element.[9]

In this report it is assumed that of mercury leaving the transfer compartment after having been inhaled, or ingested as an inorganic compound of the element or as metallic mercury, 0.08 is translocated to the kidneys and 0.92 is distributed uniformly throughout all other organs and tissues of the body. Of this mercury, whether translocated to the kidney or to any other organ or tissue, 0.95 is assumed to be retained with a biological half-life of 40 days and 0.05 with a biological half-life of 10 000 days.

## 2.2 Organic Compounds

(a) *Uptake to blood*

The absorption of methyl mercury from the gastrointestinal tract has been reviewed by Nordberg and Sherfving,[2] while more recent Japanese studies have been summarized by Kojima and Fujita.[10] These reviews suggest that methyl mercury is almost completely absorbed from the gastrointestinal tract, that the fractional absorption of mercuric acetate is about 0.2 and that the fractional absorption of phenyl mercury salts is typically 0.4.

In this report $f_1$ is taken as 1 for methyl mercury and 0.4 for other organic compounds of the element.

(b) *Inhalation classes*

There are few data concerning the inhalation of organic mercury compounds.[2] However, experiments on mice[2] suggest that di-methyl mercury should be assigned to inhalation class D. In this report all organic compounds of mercury are assigned to inhalation class D. In the absence of any relevant experimental data $f_1$ is taken as 1 for mercury entering the gastrointestinal tract as a result of the inhalation of an organic compound of the element.

| Inhalation Class | $f_1$ |
|---|---|
| D | 1 |
| W | — |
| Y | — |

(c) *Distribution and retention*

The distribution of organic compounds of mercury in the body is very dependent upon the compound administered and upon the species used (c.f. Reference 2). In the case of methyl the kidneys and brain exhibit similar high concentrations of mercury in all species studied except the rat.[2]

Experiments involving the oral administration of methyl mercury to a considerable number of individuals[8] have shown that over the first 250 days after ingestion, mercury administered in this form has a biological half-life in the whole body of $76 \pm 3$ days. However, as pointed out above, this half-life is not compatible with the total body content and daily intake of mercury given for Reference Man.[1] Nor is it compatible with the observation that mercury is still detectable in urine 6 years after the last known exposure to the element.[9]

In this report it is assumed that of mercury leaving the transfer compartment after having been inhaled or ingested as an organic compound of the element fractions 0.08 and 0.2 are translocated to the kidneys and brain respectively. The remaining fraction of mercury leaving the transfer compartment is assumed to be uniformly distributed throughout all other organs and tissues of the body. Of mercury translocated to any organ or tissue, including the kidney and brain, fractions 0.95 and 0.05 are assumed to be retained with biological half-lives of 80 and 10 000 days respectively.

It should be noted that this distribution and retention model for organic compounds of mercury is based on data for methyl mercury and is probably unduly conservative for many organic compounds of the element. For example, mercury administered as neohydrin is much more rapidly lost from the body than is methyl mercury.[11, 12]

## 3. Classification of Isotopes for Bone Dosimetry

Mercury is assumed to be uniformly distributed throughout all organs and tissues of the body other than the kidneys and brain. Therefore, a classification of isotopes of the element for the purpose of bone dosimetry is not required.

## References

1. *ICRP Publication 23, Report of the ICRP Task Group on Reference Man.* Pergamon Press, Oxford, 1975.
2. Nordberg, G. F. and Sherfving, S. Metabolism. In: *Mercury in the Environment.* Eds. Friberg, L. T. and Vostal, J. J. CRC, 1972.
3. Bornmann, G., Henke, G., Alfes, H. and Möllman, H. (1970). Uber die Enterlae Resorption von Metallischem Quecksilber. *Arch. für Toxikol.* 203–209.
4. Hursh, J. B., Clarkson, T. W., Cherian, M. G., Vostal, J. V. and Mallie, R. V. (1976). Clearance of Mercury (Hg-197, Hg-203) Vapour Inhaled by Human Subjects. *Arch. Env. Health.* **31**, 302–309.
5. ICRP Task Group on Lung Dynamics (1966). Deposition and retention models for internal dosimetry of the human respiratory tract. *Health Phys.* **12**, 173–207.
6. Morrow, P. E., Gibb, F. R. and Johnson, L. (1964). Clearance of insoluble dust from the lower respiratory tract. *Health Phys.* **10**, 543–555.
7. Kosta, L., Byrne, A. R. and Zelenko, V. (1975). Correlation between Selenium and Mercury in Man following Exposure to Inorganic Mercury. *Nature* **254**, 238–239.

8. Rahola, T., Aaran, R. K. and Miettinen, J. K. Half-life Studies of Mercury and Cadmium by Whole-body Counting. In: *Assessment of Radioactive Contamination in Man.* IAEA, Vienna, 1972, pp. 553–562.
9. Goldwater, L. J. and Nicolau, A. (1966). Absorption and Excretion of Mercury in Man: IX Persistence of Mercury in Blood and Urine following Cessation of Exposure. *Arch. Environ. Health* **12**, 196–198.
10. Kojima, K. and Fujita, M. (1973). Summary of recent studies in Japan on methyl mercury poisoning. *Toxicology* **1**, 43–62.
11. Johnson, J. E. and Johnson, J. A. (1968). A New value for the Long Component of the Effective Half-Retention Time of $^{203}$Hg in the Human. *Health Phys.* **14**, 265–266.
12. Greenlaw, κ. H. and Quaife, M. (1962). Retention of Neohydrin-Hg$^{203}$ as Determined with a Total Body Scintillation Counter. *Radiology* **78**, 970–973.
13. Newton, D. and Fry, F. A. (1978). The retention and distribution of radioactive mercuric oxide following accidental inhalation. *Ann. Occup. Hyg.* **21**, 21–32.

Annual limits on intake, ALI(Bq) and derived air concentrations, DAC(Bq/m$^3$) (40 h/wk) for isotopes of mercury

ORGANIC

| Radionuclide | | Oral | | Inhalation Class D |
|---|---|---|---|---|
| | | $f_1 = 1$ | $f_1 = 4 \times 10^{-1}$ | $f_1 = 1$ |
| $^{193m}$Hg | ALI | $3 \times 10^8$ | $2 \times 10^8$ | $5 \times 10^8$ |
| | DAC | — | — | $2 \times 10^5$ |
| $^{193}$Hg | ALI | $2 \times 10^9$ | $7 \times 10^8$ | $2 \times 10^9$ |
| | DAC | — | — | $1 \times 10^6$ |
| $^{194}$Hg | ALI | $6 \times 10^5$ | $2 \times 10^6$ | $1 \times 10^6$ |
| | DAC | — | — | $4 \times 10^2$ |
| $^{195m}$Hg | ALI | $2 \times 10^8$ | $1 \times 10^8$ | $2 \times 10^8$ |
| | DAC | — | — | $9 \times 10^4$ |
| $^{195}$Hg | ALI | $1 \times 10^9$ | $6 \times 10^8$ | $2 \times 10^9$ |
| | DAC | — | — | $7 \times 10^5$ |
| $^{197m}$Hg | ALI | $3 \times 10^8$ | $1 \times 10^8$ | $3 \times 10^8$ |
| | DAC | — | — | $1 \times 10^5$ |
| $^{197}$Hg | ALI | $4 \times 10^8$ | $3 \times 10^8$ | $5 \times 10^8$ |
| | DAC | — | — | $2 \times 10^5$ |
| $^{199m}$Hg | ALI | $2 \times 10^9$ $(4 \times 10^9)$ ST Wall | $2 \times 10^9$ $(2 \times 10^9)$ ST Wall | $6 \times 10^9$ |
| | DAC | — | — | $2 \times 10^6$ |
| $^{203}$Hg | ALI | $2 \times 10^7$ | $3 \times 10^7$ | $3 \times 10^7$ |
| | DAC | — | — | $1 \times 10^4$ |

Annual limits on intake, ALI(Bq) and derived air concentrations, DAC(Bq/m$^3$) (40 h/wk) for isotopes of mercury

### INORGANIC

| Radionuclide | | Oral | Inhalation | |
|---|---|---|---|---|
| | | | Class D | Class W |
| | | $f_1 = 2 \times 10^{-2}$ | $f_1 = 2 \times 10^{-2}$ | $f_1 = 2 \times 10^{-2}$ |
| $^{193m}$Hg | ALI | $1 \times 10^8$ | $3 \times 10^8$ | $3 \times 10^8$ |
| | DAC | — | $1 \times 10^5$ | $1 \times 10^5$ |
| $^{193}$Hg | ALI | $6 \times 10^8$ | $2 \times 10^9$ | $2 \times 10^9$ |
| | DAC | — | $7 \times 10^5$ | $6 \times 10^5$ |
| $^{194}$Hg | ALI | $3 \times 10^7$ | $2 \times 10^6$ | $4 \times 10^6$ |
| | DAC | — | $7 \times 10^2$ | $2 \times 10^3$ |
| $^{195m}$Hg | ALI | $9 \times 10^7$ | $2 \times 10^8$ | $1 \times 10^8$ |
| | DAC | — | $8 \times 10^4$ | $6 \times 10^4$ |
| $^{195}$Hg | ALI | $5 \times 10^8$ | $1 \times 10^9$ | $1 \times 10^9$ |
| | DAC | — | $5 \times 10^5$ | $5 \times 10^5$ |
| $^{197m}$Hg | ALI | $1 \times 10^8$ | $3 \times 10^8$ | $2 \times 10^8$ |
| | DAC | — | $1 \times 10^5$ | $8 \times 10^4$ |
| $^{197}$Hg | ALI | $2 \times 10^8$ | $4 \times 10^8$ | $3 \times 10^8$ |
| | DAC | — | $2 \times 10^5$ | $1 \times 10^5$ |
| $^{199m}$Hg | ALI | $2 \times 10^9$ | $5 \times 10^9$ | $7 \times 10^9$ |
| | DAC | — | $2 \times 10^6$ | $3 \times 10^6$ |
| $^{203}$Hg | ALI | $9 \times 10^7$ | $5 \times 10^7$ | $4 \times 10^7$ |
| | DAC | — | $2 \times 10^4$ | $2 \times 10^4$ |

Annual limits on intake, ALI(Bq) and derived air concentrations, DAC(Bq/m$^3$) (40 h/wk) for isotopes of mercury

### VAPOURS

| Radionuclide | | Inhalation |
|---|---|---|
| $^{193m}$Hg | ALI | $3 \times 10^8$ |
| | DAC | $1 \times 10^5$ |
| $^{193}$Hg | ALI | $1 \times 10^9$ |
| | DAC | $5 \times 10^5$ |
| $^{194}$Hg | ALI | $1 \times 10^6$ |
| | DAC | $5 \times 10^2$ |
| $^{195m}$Hg | ALI | $1 \times 10^8$ |
| | DAC | $6 \times 10^4$ |
| $^{195}$Hg | ALI | $1 \times 10^9$ |
| | DAC | $5 \times 10^5$ |
| $^{197m}$Hg | ALI | $2 \times 10^8$ |
| | DAC | $8 \times 10^4$ |
| $^{197}$Hg | ALI | $3 \times 10^8$ |
| | DAC | $1 \times 10^5$ |
| $^{199m}$Hg | ALI | $3 \times 10^9$ |
| | DAC | $1 \times 10^6$ |
| $^{203}$Hg | ALI | $3 \times 10^7$ |
| | DAC | $1 \times 10^4$ |

# METABOLIC DATA FOR LEAD

## 1. Metabolism

Data from Reference Man[1]

| | |
|---|---|
| Lead content of the body | 120 mg |
| of the skeleton | 110 mg |
| Daily intake in food and fluids | 0.44 mg |

## 2. Metabolic Model

(a) *Uptake to blood*

The fractional absorption of lead from the gastrointestinal tract of man has been estimated to be in the range 0.05 to 0.14.[1] More recently, studies by Chamberlain and his co-workers[2] have demonstrated considerably more variation in the fractional absorption of lead compounds. In particular, after fasting the gastrointestinal absorption of $PbCl_2$ ranged from 0.24 to 0.65 in six normal subjects. Similarly, the studies of Wetherill *et al.*[3] and Blake[4] also indicate values for fractional gastrointestinal absorption considerably greater than 0.1. In this report $f_1$ is taken to be 0.2, a value thought to be appropriate for compounds of lead ingested between meals.[2,4]

(b) *Inhalation classes*

The ICRP Task Group on Lung Dynamics[5] assigned all commonly occurring compounds of lead to inhalation class W. However, various studies have indicated that lead aerosols with submicron AMADs are rapidly cleared from the lungs.[2,6-9] In this report all commonly occurring compounds of lead are assigned to inhalation class D.

| Inhalation Class | $f_1$ |
|---|---|
| D | 0.2 |
| W | — |
| Y | — |

(c) *Distribution and retention*

Lloyd *et al.*[10] have shown that the whole body retention of lead in dogs is well described by the function.

$$R(t) = 0.7e^{-0.693t/12} + 0.17e^{-0.693t/180} + 0.13e^{-0.693t/5000}$$

although the half-life of the third component was not well defined by their experiments. A long-term component of lead retention with a biological half-life of about 10000 days is also indicated in man.[11]

Injected $^{210}Pb$ is primarily deposited in bone, liver and kidneys but is tenaciously retained only by mineral bone.[10,12] From the data given by Lloyd *et al.*[10] for the tissue distribution of $^{210}Pb$ at 28, 1100 and 1497 days post-injection, the distribution of stable lead in Reference Man[1] and the whole-body retention data in dogs[10] the following metabolic model has been adopted.

Of lead leaving the transfer compartment fractions 0.55, 0.25, and 0.02 are assumed to be translocated to the skeleton, liver and kidneys respectively. The remaining fraction, 0.18, is assumed to be uniformly distributed throughout all other organs and tissues of the body. Of lead translocated to the skeleton fractions 0.6, 0.2 and 0.2 are assumed to be retained with biological half-lives of 12, 180 and 10 000 days respectively. Of lead translocated to any organ or tissue other than the skeleton fractions 0.8, 0.18 and 0.02 are assumed to be retained with biological half-lives of 12, 180 and 10 000 days respectively.

### 3. Classification of Isotopes for Bone Dosimetry

It is known that lead can be substituted in the calcium positions of apatite[13] and it is probable, therefore, that the element distributes fairly rapidly throughout the volume of mineral bone. In this report it is assumed that $^{202}$Pb, $^{205}$Pb and $^{210}$Pb are uniformly distributed throughout the volume of mineral bone at all times after their deposition in the skeleton. All other isotopes of lead considered in this report have radioactive half-lives of less than 15 days and are assumed to be uniformly distributed on bone surfaces at all times following their deposition in the skeleton.

### References

1. *ICRP Publication 23, Report of the ICRP Task Group on Reference Man.* Pergamon Press, Oxford, 1975.
2. Chamberlain, A. C., Heard, M. J., Little, P., Newton, D., Wells, A. C. and Wiffen, R. D. *Investigations into lead from motor vehicles.* United Kingdom Atomic Energy Authority (Harwell) report AERE-R 9198 H.M. Stationery Office, London, 1978.
3. Wetherill, G. W., Rabinowitz, M. and Kopple, J. D. Sources and metabolic pathways of lead in normal humans. In: *Recent advances in the assessment of the health effects of environmental pollution.* CEC, Paris, 1975, pp. 847–860.
4. Blake, K. C. H. (1976). Absorption of $^{203}$Pb from the gastro-intestinal tract of man. *Environ. Res.* **11**, 1–4.
5. ICRP Task Group on Lung Dynamics (1966). Deposition and retention models for internal dosimetry of the human respiratory tract. *Health Phys.*, **12**, 173–207.
6. Bianco, A., Gibb, F. R. and Morrow, P. E. *Study of submicron size lead-212 aerosol inhalation.* Abstracts of papers presented at the third international congress of the International Radiation Protection Association, Washington. Pergamon Press, Oxford, 1973.
7. Hursh, J. B., Schraub, A., Sattler, E. L. and Hofmann, H. P. (1969). Fate of $^{212}$Pb Inhaled by Human Subjects. *Health Phys.* **16**, 257–267.
8. Booker, D. V., Chamberlain, A. C., Newton, D. and Stott, A. N. B. (1969). Uptake of radioactive lead following inhalation and injection. *Br. J. Radiol.* **42**, 457–466.
9. Hursh, J. B. and Mercer, T. T. (1970). Measurement of $^{212}$Pb loss rate from human lungs. *J. Appl. Physiol.* **28**, 268–274.
10. Lloyd, R. D., Mays, C. W., Atherton, D. R. and Bruenger, F. W. (1975). $^{210}$Pb Studies in Beagles. *Health Phys.* **28**, 575–583.
11. Rabinowitz, M. B., Wetherill, G. W. and Kopple, J. D. (1973). Lead Metabolism in the Normal Human: Stable Isotope Studies. *Science,* **182**, 725–727.
12. Cohen, N. The Retention and Distribution of Lead-210 in the Adult Baboon. *Annual Progress Report, Sept. 1, 1969–August 31, 1970.* NYO-3086-10, vol. 1, New York University, N.Y., Institute of Environmental Medicine.
13. Posner, A. S. Mineralised Tissues. In: *Phosphorus and its Compounds,* Vol. 2, Ed. Van Wazer, J. R. (Interscience, New York, 1961), pp. 1429–1459.

Annual limits on intake, ALI(Bq) and derived air concentrations, DAC(Bq/m$^3$) (40 h/wk) for isotopes of lead

| Radionuclide | | Oral | Inhalation Class D |
|---|---|---|---|
| | | $f_1 = 2 \times 10^{-1}$ | $f_1 = 2 \times 10^{-1}$ |
| $^{195m}$Pb | ALI | $2 \times 10^9$ | $7 \times 10^9$ |
| | DAC | — | $3 \times 10^6$ |
| $^{198}$Pb | ALI | $1 \times 10^9$ | $2 \times 10^9$ |
| | DAC | — | $1 \times 10^6$ |
| $^{199}$Pb | ALI | $8 \times 10^8$ | $3 \times 10^9$ |
| | DAC | — | $1 \times 10^6$ |
| $^{200}$Pb | ALI | $1 \times 10^8$ | $2 \times 10^8$ |
| | DAC | — | $1 \times 10^5$ |
| $^{201}$Pb | ALI | $3 \times 10^8$ | $7 \times 10^8$ |
| | DAC | — | $3 \times 10^5$ |
| $^{202m}$Pb | ALI | $3 \times 10^8$ | $1 \times 10^9$ |
| | DAC | — | $4 \times 10^5$ |
| $^{202}$Pb | ALI | $5 \times 10^6$ | $2 \times 10^6$ |
| | DAC | — | $8 \times 10^2$ |
| $^{203}$Pb | ALI | $2 \times 10^8$ | $4 \times 10^8$ |
| | DAC | — | $1 \times 10^5$ |
| $^{205}$Pb | ALI | $1 \times 10^8$ | $5 \times 10^7$ |
| | DAC | — | $2 \times 10^4$ |
| $^{209}$Pb | ALI | $9 \times 10^8$ | $2 \times 10^9$ |
| | DAC | — | $9 \times 10^5$ |
| $^{210}$Pb | ALI | $2 \times 10^4$ $(4 \times 10^4)$ Bone surf. | $9 \times 10^3$ $(1 \times 10^4)$ Bone surf. |
| | DAC | — | 4 |
| $^{211}$Pb | ALI | $4 \times 10^8$ | $2 \times 10^7$ |
| | DAC | — | $1 \times 10^4$ |
| $^{212}$Pb | ALI | $3 \times 10^6$ $(5 \times 10^6)$ Bone surf. | $1 \times 10^6$ |
| | DAC | — | $5 \times 10^2$ |
| $^{214}$Pb | ALI | $3 \times 10^8$ | $3 \times 10^7$ |
| | DAC | — | $1 \times 10^4$ |

# METABOLIC DATA FOR BISMUTH

## 1. Metabolism

Data from Reference Man[1]
Bismuth content of soft tissues      <0.23 mg
Daily intake in food and fluids      20 $\mu$g

## 2. Metabolic Model

(a) *Uptake to blood*

It has been suggested that the fractional absorption of dietary bismuth from the gastrointestinal tract is about 0.08.[1] It has also been stated[2] that basic salts of bismuth are only poorly absorbed from the gastrointestinal tract. In this report $f_1$ is taken to be 0.05, a value which is thought to be appropriate for those compounds of the element most likely to be encountered in practice.

(b) *Inhalation classes*

There appears to be no relevant experimental information available concerning the inhalation of compounds of bismuth. The ICRP Task Group on Lung Dynamics[3] assigned bismuth nitrate to inhalation class D and all other compounds of the element to inhalation class W. In the absence of any relevant experimental data this classification is adopted here.

| Inhalation Class | $f_1$ |
|---|---|
| D | 0.05 |
| W | 0.05 |
| Y | — |

(c) *Distribution and retention*

Bismuth is primarily deposited in the kidney[4-7] the concentrations in spleen, bone, liver and lung being more than an order of magnitude lower than the concentration found in the kidney.

In this report it is assumed that of bismuth leaving the transfer compartment 0.3 goes directly to excreta, 0.4 is translocated to the kidneys and 0.3 is uniformly distributed throughout all other organs and tissues of the body.[6,7] Of bismuth translocated to any organ or tissue of the body, including the kidneys, fractions 0.6 and 0.4 are assumed to be retained with biological half-lives of 0.6 and 5 days respectively.[6,7]

Since bismuth is very rapidly cleared from the blood[7] the half-life of clearance from the transfer compartment is assumed to be 0.01 day for all compounds of bismuth.

## 3. Classification of Isotopes for Bone Dosimetry

Systemic bismuth is assumed to be uniformly distributed throughout all organs and tissues of the body other than the kidneys. Therefore, a classification of isotopes of the element for the purpose of bone dosimetry is not required.

# References

1. *ICRP Publication 23, Report of the ICRP Task Group on Reference Man.* Pergamon Press, Oxford, 1975.
2. Sollmann, T. *A Manual of Pharmacology.* W. B. Saunders, Philadelphia, 1957, pp. 1237–1245.
3. ICRP Task Group on Lung Dynamics (1966). Deposition and retention models for the internal dosimetry of the human respiratory tract. *Health Phys.* **12**, 173–207.
4. Eridani, S., Balzarini, M., Taglioretti, D., Romussi, M. and Velentini, R. (1964). The Distribution of Radiobismuth in the Rat. *Br. J. Radiol.* **37**, 311–314.
5. Durbin, P. W. (1960). Metabolic Characteristics within a Chemical Family. *Health Phys.* **2**, 225–238.
6. Matthews, C. M. E., Dempster, W. J., Kapros, C. and Kountz, S. (1964). The Effect of Bismuth 206 Irradiation on Survival of Skin Homografts. *Br. J. Radiol.* **37**, 306–310.
7. Russ, G. A., Bigler, R. E., Tilbury, R. S., Woodward, H. Q. and Laughlin, S. (1975). Metabolic Studies with Radiobismuth 1. Retention and Distribution of $^{206}$Bi in the Normal Rat. *Rad. Res.* **63**, 443–454.

Annual limits on intake, ALI(Bq) and derived air concentrations, DAC(Bq/m$^3$)
(40 h/wk) for isotopes of bismuth

| Radionuclide | | Oral | Inhalation | |
|---|---|---|---|---|
| | | | Class D | Class W |
| | | $f_1 = 5 \times 10^{-2}$ | $f_1 = 5 \times 10^{-2}$ | $f_1 = 5 \times 10^{-2}$ |
| $^{200}$Bi | ALI | $1 \times 10^9$ | $3 \times 10^9$ | $4 \times 10^9$ |
| | DAC | — | $1 \times 10^6$ | $2 \times 10^6$ |
| $^{201}$Bi | ALI | $4 \times 10^8$ | $1 \times 10^9$ | $1 \times 10^9$ |
| | DAC | — | $4 \times 10^5$ | $6 \times 10^5$ |
| $^{202}$Bi | ALI | $5 \times 10^8$ | $1 \times 10^9$ | $3 \times 10^9$ |
| | DAC | — | $6 \times 10^5$ | $1 \times 10^6$ |
| $^{203}$Bi | ALI | $9 \times 10^7$ | $2 \times 10^8$ | $2 \times 10^8$ |
| | DAC | — | $1 \times 10^5$ | $9 \times 10^4$ |
| $^{205}$Bi | ALI | $5 \times 10^7$ | $9 \times 10^7$ | $5 \times 10^7$ |
| | DAC | — | $4 \times 10^4$ | $2 \times 10^4$ |
| $^{206}$Bi | ALI | $2 \times 10^7$ | $5 \times 10^7$ | $3 \times 10^7$ |
| | DAC | — | $2 \times 10^4$ | $1 \times 10^4$ |
| $^{207}$Bi | ALI | $4 \times 10^7$ | $6 \times 10^7$ | $1 \times 10^7$ |
| | DAC | — | $3 \times 10^4$ | $5 \times 10^3$ |
| $^{210m}$Bi | ALI | $2 \times 10^6$ $(2 \times 10^6)$ Kidneys | $2 \times 10^5$ $(2 \times 10^5)$ Kidneys | $3 \times 10^4$ |
| | DAC | — | $7 \times 10^1$ | $1 \times 10^1$ |
| $^{210}$Bi | ALI | $3 \times 10^7$ | $9 \times 10^6$ $(1 \times 10^7)$ Kidneys | $1 \times 10^6$ |
| | DAC | — | $4 \times 10^3$ | $4 \times 10^2$ |
| $^{212}$Bi | ALI | $2 \times 10^8$ | $9 \times 10^6$ | $1 \times 10^7$ |
| | DAC | — | $4 \times 10^3$ | $4 \times 10^3$ |
| $^{213}$Bi | ALI | $3 \times 10^8$ | $1 \times 10^7$ | $1 \times 10^7$ |
| | DAC | — | $5 \times 10^3$ | $5 \times 10^3$ |
| $^{214}$Bi | ALI | $6 \times 10^8$ $(8 \times 10^8)$ ST WALL | $3 \times 10^7$ | $3 \times 10^7$ |
| | DAC | — | $1 \times 10^4$ | $1 \times 10^4$ |

# METABOLIC DATA FOR NEPTUNIUM

## 1. Metabolism

Neptunium is not a naturally occurring element and data have not been obtained on its distribution in man. Reliance must, therefore be placed on animal data. These data have been reviewed by an ICRP Task Group.[1]

## 2. Metabolic Model

(a) *Uptake to blood*

Experiments on rats[2-4] indicate that the fractional absorption of neptunium from the gastrointestinal tract is about 0.01 when it is administered as the nitrate. However, the fractional absorption of trace quantities of the element may be a factor of ten lower[2] as may be the fractional absorption of neptunium incorporated in food.[4] In this report $f_1$ is taken to be 0.01 for all compounds of neptunium.

(b) *Inhalation classes*

Experiments on rats[5-7] indicate that neptunium is cleared from the lung more rapidly than plutonium. From the limited rodent data and by analogy with americium, all compounds of the element have been assigned to inhalation class W.

| Inhalation Class | $f_1$ |
| --- | --- |
| D | — |
| W | 0.01 |
| Y | — |

(c) *Distribution and retention*

Data on the distribution and retention of neptunium in the rat[2-4,6-10] indicate that its metabolic behaviour is rather similar to that of plutonium. However, there are some indications that in the skeleton neptunium may distribute more like calcium than like plutonium.[11] In this report the metabolic model used for plutonium has also been used for neptunium in accordance with the Task Group's recommendations.[1] Thus, of neptunium leaving the transfer compartment 0.45 is assumed to be translocated to mineral bone and 0.45 to the liver. The fraction of neptunium translocated to the gonads is assumed to be $3.5 \times 10^{-4}$ for the testes and $1.1 \times 10^{-4}$ for the ovaries, these values corresponding to a fractional translocation to the gonads of $10^{-5}$ per gram of gonadal tissue. The remainder of neptunium leaving the transfer compartment is assumed to go directly to excreta.

Neptunium translocated to mineral bone is assumed to be retained in that tissue with a biological half-life of 100 years, while neptunium translocated to the liver is assumed to be retained in that tissue with a biological half-life of 40 years. Neptunium translocated to the gonads is assumed to be permanently retained in that tissue.

## 3. Classification of Isotopes for Bone Dosimetry

The initial deposition of neptunium in bone is more like calcium than plutonium.[11] It is, nevertheless, still primarily associated with bone surfaces at early times after deposition and is unlikely to be as mobile in the bone matrix as are the alkaline earths. For these reasons neptunium is, like the other actinides, assumed to be uniformly distributed over the endosteal surfaces of mineral bone at all times following its deposition in the skeleton.

### References

1. *ICRP Publication 19, The Metabolism of Compounds of Plutonium and Other Actinides.* Pergamon Press, Oxford, 1972.
2. Ballou, J. E., Bair, W. J., Case, A. C. and Thompson, R. C. (1962). Studies with Neptunium in the Rat. *Health Phys.* **8**, 685–688.
3. Sullivan, M. F. and Crosby, A. L. Absorption of Uranium-233, Neptunium-237, Plutonium-238, Americium-241, Curium-244, and Einsteinium-253 from the Gastrointestinal Tract of Newborn and Adult Rats. *Battelle Pacific Northwest Laboratories Annual Report for 1974, Part 1.* BNWL-1950 PT1, 1975, pp. 105–108.
4. Sullivan, M. F. and Crosby, A. L. Absorption of Transuranic Elements from Rat Gut. *Battelle Pacific Northwest Laboratories Annual Report for 1975, Part 1.* BNWL-2000-PT1, 1976, pp. 91–93.
5. Bair, W. J. and Case, A. C. Preliminary Studies of Inhaled Dust Containing Neptunium-237. *AEC Research and Development Report* HW-70949, 1961.
6. Lyubchanskii, E. R. and Levdik, T. I. Neptunium 237 Metabolism after Inhalation thereof. In: *Biologicheskoye Deystviye Vneshnikh i Vnutrennikh Istochnikov Radiatsii.* Eds. Moskalev, Yu. I. and Kalistratova, V. S., (Meditsina, Moscow, 1972) pp. 204–214, translated in AEC-tr-7457 (1972) pp. 309–321.
7. Moskalev, Yu. I., Rudnitskaya, E. I., Zalikin, G. A., Petrovich, I. K. and Levdik, T. I. In: *Proceedings of German-Soviet Working Meeting on the Question of Radiation Protection.* Berlin (1972), SZS-148 (1973) pp. 12–28.
8. Moskalev, Yu. I., Rudnitskaya, E. N., Zalikin, G. A. and Petrovich, I. K. Distribution and biological effects of neptunium 237. In: *Biologicheskoye Deystviye Vneshnikh i Vnutrennikh Istochnikov Radiatsii.* Eds. Moskalev, Yu. I. and Kalistratova, V. S. Meditsina, Moscow, 1972, pp. 220–229, translated in AEC-tr-7457 (1972) pp. 330–341.
9. Levdik, T. I., Lemberg, V. K., Yerokhin, R. A. and Buldakov, L. A. Some characteristics of the biological effect and behaviour of neptunium-237 in the animal body after administering different isotope salts. In: *Otdalennyyee Posledstviya Luchevykh Porazheniy.* Ed. Moskalev, Yu. I. Atomizdat, Moscow, 1971, pp. 439–449, translated in AEC-tr-7387 (1972), pp. 483–494.
10. Mahlum, D. D. and Clarke, W. J. (1966). Neptunium-237 Toxicity in the Rat-I. Histopathologic and Chemical Observations in Liver and Kidney. *Health Phys.* **12**, 7–13.
11. Nenot, J. C., Masse, R., Morin, M. and Lafuma, J. (1972). An Experimental Comparative Study of the Behaviour of $^{237}$Np, $^{238}$Pu, $^{241}$Am and $^{242}$Cm in Bone. *Health Phys.* **22**, 657–666.

Annual limits on intake, ALI(Bq) and derived air concentrations,
DAC(Bq/m$^3$) (40 h/wk) for isotopes of neptunium

| Radionuclide | | Oral | Inhalation Class W |
|---|---|---|---|
| | | $f_1 = 1 \times 10^{-2}$ | $f_1 = 1 \times 10^{-2}$ |
| $^{232}$Np | ALI | $1 \times 10^9$ $(2 \times 10^9)$ Bone surf. | $9 \times 10^7$ $(2 \times 10^8)$ Bone surf. |
| | DAC | — | $4 \times 10^4$ |
| $^{233}$Np | ALI | $3 \times 10^{10}$ | $1 \times 10^{11}$ |
| | DAC | — | $5 \times 10^7$ |
| $^{234}$Np | ALI | $8 \times 10^7$ | $1 \times 10^8$ |
| | DAC | — | $4 \times 10^4$ |
| $^{235}$Np | ALI | $4 \times 10^8$ | $5 \times 10^7$ $(5 \times 10^7)$ Bone surf. |
| | DAC | --- | $2 \times 10^4$ |
| $^{236}$Np $(1.15 \times 10^5 \text{y})$ | ALI | $1 \times 10^4$ $(2 \times 10^4)$ Bone surf. | $1 \times 10^3$ $(2 \times 10^3)$ Bone surf. |
| | DAC | — | $4 \times 10^{-1}$ |
| $^{236}$Np (22.5 h) | ALI | $2 \times 10^7$ $(3 \times 10^7)$ Bone surf. | $1 \times 10^6$ $(3 \times 10^6)$ Bone surf. |
| | DAC | — | $6 \times 10^2$ |
| $^{237}$Np | ALI | $3 \times 10^3$ $(5 \times 10^3)$ Bone surf. | $2 \times 10^2$ $(4 \times 10^2)$ Bone surf. |
| | DAC | · — | $9 \times 10^{-2}$ |
| $^{238}$Np | ALI | $3 \times 10^7$ | $3 \times 10^6$ $(6 \times 10^6)$ Bone surf. |
| | DAC | — | $1 \times 10^3$ |
| $^{239}$Np | ALI | $6 \times 10^7$ | $9 \times 10^7$ |
| | DAC | $(6 \times 10^7)$ LLI Wall | |
| | DAC | — | $4 \times 10^4$ |
| $^{240}$Np | ALI | $8 \times 10^8$ | $3 \times 10^9$ |
| | DAC | — | $1 \times 10^6$ |